MAKERS

MAKERS

The New Industrial Revolution

CHRIS ANDERSON

CROWN
BUSINESS
NEW YORK

Originally published in hardcover in the United States by Crown Business,
an imprint of the Crown Publishing Group, a division of Random House LLC,
New York, in 2012.

Crown Business books are available at special discounts for bulk purchases for
sales promotions or corporate use. Special editions, including personalized covers,
excerpts of existing books, or books with corporate logos, can be created in large
quantities for special needs. For more information, contact Premium Sales at
(212) 572-2232 or e-mail specialmarkets@randomhouse.com.

Library of Congress Cataloging-in-Publication Data

Anderson, Chris.
Makers : the new industrial revolution / Chris Anderson.
        p.   cm.
Includes bibliographical references and index.
    1. Entrepreneurship.   2. Microfabrication.   3. Micromachining.
4. Business enterprises—Technological innovations.   I. Title.
    HB615.A683 2013
    338'.04—dc23   2012014398

ISBN 978-0-307-72096-2
eBook ISBN 978-0-307-72097-9

Printed in the United States of America

Book design by Lauren Dong
Creative direction and design by Brandon Kavulla

10 9 8 7 6 5 4 3 2 1

First Paperback Edition

For Carlotta Anderson

Contents

MAKERS

Part One

The Revolution

Chapter 1

The Invention Revolution

Fred Hauser, my maternal grandfather, emigrated to Los Angeles from Bern, Switzerland, in 1926. He was trained as a machinist, and perhaps inevitably for Swiss mechanical types, there was a bit of the watchmaker in him, too. Fortunately, at that time the young Hollywood was something of a clockwork industry, too, with its mechanical cameras, projection systems, and the new technology of magnetic audio strips. Hauser got a job at MGM Studios working on recording technology, got married, had a daughter (my mom), and settled in a Mediterranean bungalow on a side street in Westwood where every house had a lush front lawn and a garage in the back.

But Hauser was more than a company engineer. By night, he was also an inventor. He dreamed of machines, drew sketches and then mechanical drawings of them, and built prototypes. He converted his garage to a workshop, and gradually equipped it with the tools of creation: a drill press, a band saw, a jig saw, grinders, and, most important, a full-size metal lathe, which is a miraculous device that can, in the hands of an expert operator, turn blocks of steel or aluminum into precision-machined mechanical sculpture ranging from camshafts to valves.

Initially his inventions were inspired by his day job, and involved various kinds of tape-transport mechanisms. But over time his attention shifted to the front lawn. The hot California sun and the local mania for perfect green-grass plots had led to a booming industry in sprinkler systems, and as the region grew prosperous, gardens were torn up to lay irrigation systems. Proud homeowners came home from work, turned on the valves, and admired the water-powered wizardry of pop-up rotors, variable-stream nozzles, and impact sprinkler heads spreading water beautifully around their plots. Impressive, aside from the fact that they all required manual intervention, if nothing more than just to turn on the valves in the first place. What if they could be driven by some kind of clockwork, too?

Patent number 2311108 for "Sequential Operation of Service Valves," filed in 1943, was Hauser's answer. The patent was for an automatic sprinkler system, which was basically an electric clock that turned water valves on and off. The clever part, which you can still find echoes of today in lamp timers and thermostats, is the method of programming: the "clock" face is perforated with rings of holes along the rim at each five-minute mark. A pin placed in any hole triggers an electrical actuator called a solenoid, which toggles a water valve on or off to control that part of the sprinkler system. Each ring rep-

resented a different branch of the irrigation network. Together they could manage an entire yard—front, back, patio, and driveway areas.

Once he had constructed the prototype and tested it in his own garden, Hauser filed his patent. With the patent application pending, he sought to bring it to market. And there was where the limits of the twentieth-century industrial model were revealed.

It used to be hard to change the world with an idea alone. You can invent a better mousetrap, but if you can't make it in the millions, the world won't beat a path to your door. As Marx observed, power belongs to those who control the means of production. My grandfather could invent the automatic sprinkler system in his workshop, but he couldn't build a factory there. To get to market, he had to interest a manufacturer in licensing his invention. And that is not only hard,

but requires the inventor to lose control of his or her invention. The owners of the means of production get to decide what is produced.

In the end, my grandfather got lucky—to a point. Southern California was the center of the new home irrigation industry, and after much pitching, a company called Moody agreed to license his automatic sprinkler system. In 1950 it reached the market as the Moody Rainmaster, with a promise to liberate homeowners so they could go to the beach for the weekend while their gardens watered themselves. It sold well, and was followed by increasingly sophisticated designs, for which my grandfather was paid royalties until the last of his automatic sprinkler patents expired in the 1970s.

This was a one-in-a-thousand success story; most inventors toil in their workshops and never get to market. But despite at least twenty-six other patents on other devices, he never had another commercial hit. By the time he died in 1988, I estimate he had earned only a few hundred thousand dollars in total royalties. I remember visiting the company that later bought Moody, Hydro-Rain, with him as a child in the 1970s to see his final sprinkler system model being made. They called him "Mr. Hauser" and were respectful, but it was apparent they didn't know why he was there. Once they had licensed the patents, they then engineered their own sprinkler systems, designed to be manufacturable, economical, and attractive to the buyer's eye. They bore no more resemblance to his prototypes than his prototypes did to his earliest tabletop sketches.

This was as it must be; Hydro-Rain was a company making many tens of thousands of units of a product in a competitive market driven by price and marketing. Hauser, on the other hand, was a little old Swiss immigrant with an expiring invention claim who worked out of a converted garage. He didn't belong at the factory, and they didn't need him. I remember that some hippies in a Volkswagen yelled at him for driving too slowly on the highway back from the factory. I was twelve and mortified. If my grandfather was a hero of twentieth-century capitalism, it certainly didn't look that way. He just seemed like a tinkerer, lost in the real world.

Yet Hauser's story is no tragedy; indeed, it was a rare success story from that era. My grandfather was, as best I can remember (or was able to detect; he fit the caricature of a Swiss engineer, more comfortable with a drafting pencil than with conversation), happy, and he lived luxuriously by his standards. I suspect he was compensated relatively fairly for his patent, even if my stepgrandmother (my grandmother died early) complained about the royalty rates and his lack of aggressiveness in negotiating them. He was by any measure an accomplished inventor. But after his death, as I went through his scores of patent filings, including a clock timer for a stove and a Dictaphone-like recording machine, I couldn't help but observe that of his many ideas, only the sprinklers actually made it to market at all.

Why? Because he was an inventor, not an entrepreneur. And in that distinction lies the core of this book.

It used to be hard to be an entrepreneur. The great inventors/businessmen of the First Industrial Revolution, such as James Watt and Matthew Boulton of steam-engine fame, were not just smart but privileged. Most were either born into the ruling class or lucky enough to be apprenticed to one of the elite. For most of history since then, entrepreneurship has meant either setting up a corner grocery shop or some other sort of modest local business or, more rarely, a total pie-in-the-sky crapshoot around an idea that is more likely to bring ruination than riches.

Today we are spoiled by the easy pickings of the Web. Any kid with an idea and a laptop can create the seeds of a world-changing company—just look at Mark Zuckerberg and Facebook or any one of thousands of other Web startups hoping to follow his path. Sure, they may fail, but the cost is measured in overdue credit-card payments, not lifelong disgrace and a pauper's prison.

The beauty of the Web is that it democratized the tools both of invention *and* of production. Anyone with an idea for a service can turn it into a product with some software code (these days it hardly even requires much programming skill, and what you need you can learn online)—no patent required. Then, with a keystroke, you can "ship it" to a global market of billions of people.

Maybe lots of people will notice and like it, or maybe they won't. Maybe there will be a business model attached, or maybe there won't. Maybe riches lie at the end of this rainbow, or maybe they don't. But the point is that the path from "inventor" to "entrepreneur" is so fore-shortened it hardly exists at all anymore.

Indeed, startup factories such as Y Combinator now coin entre-preneurs first and ideas later. Their "startup schools" admit smart young people on the basis of little more than a PowerPoint presenta-tion. Once admitted, the would-be entrepreneurs are given spending money, whiteboards, and desk space and told to dream up something worth funding in three weeks.

Most do, which says as much about the Web's ankle-high barriers to entry as it does about the genius of the participants. Over the past six years, Y Combinator has funded three hundred such companies, with such names as Loopt, Wufoo, Xobni, Heroku, Heyzap, and Bump. Incredibly, some of them (such as DropBox and Airbnb) are now worth billions of dollars. Indeed, the company I work for, Condé Nast, even bought one of them, Reddit, which now gets more than 2 billion page views a month. It's on its third team of twentysome-thing genius managers; for some of them, this is their first job and they've never known anything but stratospheric professional success.

But that is the world of bits, those elemental units of the digi-tal world. The Web Age has liberated bits; they are cheaply created and travel cheaply, too. This is fantastic; the weightless economics of bits has reshaped everything from culture to economics. It is perhaps the defining characteristic of the twenty-first century (I've written a couple of books on that, too). Bits have changed the world.

We, however, live mostly in the world of atoms, also known as the Real World of Places and Stuff. Huge as information industries have be-come, they're still a sideshow in the world economy. To put a ballpark fig-ure on it, the digital economy, broadly defined, represents $20 trillion of revenues, according to Citibank and Oxford Economics.[1] The economy beyond the Web, by the same estimate, is about $130 trillion. In short, the world of atoms is at least five times larger than the world of bits.

We've seen what the Web's model of democratized innovation has done to spur entrepreneurship and economic growth. Just imagine what a similar model could do in the larger economy of Real Stuff. More to the point, there's no need to imagine—it's already starting to happen. That's what this book is about. There are thousands of entrepreneurs emerging today from the Maker Movement who are industrializing the do-it-yourself (DIY) spirit. I think my grandfather, as bemused as he might be by today's open-source and online "co-creation," would resonate with the Maker Movement. Indeed, I think he might be proud.

The making of a Maker

In the 1970s, I spent some of my happiest childhood summers with my grandfather in Los Angeles, visiting from my home on the East

Coast and learning to work with my hands in his workshop. One spring, he announced that we would be making a four-stroke gasoline engine and that he had ordered a kit we could build together. When I arrived in Los Angeles that summer, the box was waiting. I had built my share of models, and opened the box expecting the usual numbered parts and assembly instructions. Instead, there were three big blocks of metal and a crudely cast engine casing. And a large blueprint, a single sheet folded many times.

"Where are the parts?" I asked. "They're in there," my grandfather replied, pointing to the metal blocks. "It's our job to get them out." And that's exactly what we did that summer. Using the blueprint as a guide, we cut, drilled, ground, and turned those blocks of metal, extracting a crankshaft, piston and rod, bearings, and valves out of solid brass and steel, much as an artist extracts a sculpture from a block of marble. As the pile of metal curlicues from the steel turning on the lathe grew around my feet, I marveled at the power of tools

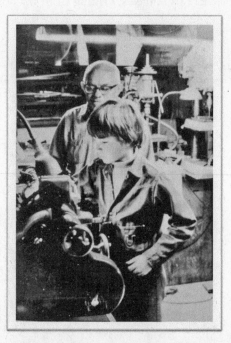

and skilled hands (my grandfather's, not mine). We had conjured a precision machine from a lump of metal. We were a mini-factory, and we could make anything.

But as I got older, I stopped returning to my grandfather's workshop and forgot about my fascination with making things. Blame screens. My generation was the first to get personal computers, and I was more enthralled with them than with anything my grandfather could make. I learned to program, and my creations were in code, not steel. Tinkering in a workshop seemed trivial compared to unlocking the power of a microprocessor.

Zines, Sex Pistols, and the birth of Indie

When I reached my twenties, I had my second DIY moment. I was living in Washington, D.C., in the early 1980s, when it was one of the hotspots of the American punk rock movement. Bands such as Minor Threat and the Teen Idles were being formed by white suburban teenagers and playing in church basements. Despite not knowing how to play an instrument and having limited talent, I got caught up in the excitement of the moment and played in some of the lesser bands in the scene.[2] It was eye-opening.

Like all garage rock and roll, all you needed to be in a band was an electric guitar and an amp. But what was new about the 1980s punk phenomenon was that the bands did more than just play; they also started to publish. Photocopiers were becoming common, and from them arose a "zine" culture of DIY magazines that were distributed at stores and shows and by mail. Cheap four-track tape recorders allowed bands to record and mix their own music, without a professional studio. And a growing industry of small vinyl-pressing plants let them make small-batch singles and EPs, which they sold via mail order and local shops.

This was the start of the DIY music industry. The tools of the major labels—recording, manufacturing, and marketing music—were now in the hands of individuals. Eventually some of these bands, led by Minor Threat and then Fugazi, started their own indie label, Dischord, which eventually produced hundreds of records and is still running today. They didn't need to compromise their music to get published, and they didn't need to sell in big numbers or get radio play. They could find their own fans; indeed, the fans found them via word of mouth, and postcards poured into such micro-labels to order music that couldn't be found in most stores. The relative obscurity conferred authenticity and contributed to the rise of the global underground that defines Web culture today.

My bands did all of this: from the photocopied flyers to the zines to the four-track tapes to the indie-label albums. We never got very big, but that wasn't the point. We still had day jobs, but we were doing what we thought was genuinely innovative and getting people at our shows, even touring to New York and to other cities with their own indie music scenes. Out of this came the roots of what would become today's alternative rock world.

By the time I was in my mid-twenties, it was clear that my talents lay elsewhere and I left music. I went back to college and, in part to make up for lost time, decided to major in the hardest subject I could find, physics. Although I wasn't terribly good at that, either, it did expose me to the beginnings of the Internet, which you'll recall started as a way for academic labs, especially big physics facilities with expensive equipment used by researchers from around the world, to connect to each other.

After graduating and working summers at some physics labs, I started working as a writer for the science journals *Nature* and *Science*, which were still part of the academic world and users of the early Internet. That in turn brought me to my third DIY chapter, the Web, which was created in 1990 at CERN, a physics laboratory in Switzerland. Once I saw that, just months after the first websites went live, I realized that I had been incredibly lucky to be in the right place at the right time. I was witnessing the birth of a new medium, one that I not only could be a part of, but could help promote.

From my start in the science world to my job today editing *Wired*, the digital revolution became my career. In the Web Age, the DIY punk movement's co-opting of the means of production turned into regular people using desktop publishing, then websites, then blogs, and now social media. Indie-pressed vinyl became YouTube music videos. Four-track tape recorders became ProTools and iPad music apps. Garage bands became Apple's GarageBand.

Now, three decades later, I find my thoughts returning to my grandfather's garage. It's not nostalgia, nor have I changed my mind about the digital revolution. It's just that the digital revolution has now reached the workshop, the lair of Real Stuff, and there it may have its greatest impact yet. Not just the workshops themselves (although they're getting pretty cool these days), but more what can be done in the physical world by regular people with extraordinary tools.

We are all Makers. We are born Makers (just watch a child's fascination with drawing, blocks, Lego, or crafts), and many of us retain that love in our hobbies and passions. It's not just about workshops, garages, and man caves. If you love to cook, you're a kitchen Maker and your stove is your workbench (homemade food is best, right?). If you love to plant, you're a garden Maker. Knitting and sewing, scrapbooking, beading, and cross-stitching—all Making.

These projects represent the ideas, dreams, and passions of millions of people. Most never leave the home, and that's probably no bad thing. But one of the most profound shifts of the Web Age is that there is a new default of sharing online. If you do something, video it. If you video something, post it. If you post something, promote it to your friends. Projects shared online become inspiration for others and opportunities for collaboration. Individual Makers, globally connected this way, become a movement. Millions of DIYers, once working alone, suddenly start working together.

Thus ideas, shared, turn into bigger ideas. Projects, shared, become group projects and more ambitious than any one person would attempt alone. And those projects can become the seeds of products, movements, even industries. The simple act of "making in public" can become the engine of innovation, even if that was not the intent. It is simply what ideas do: spread when shared.

We've seen this play out on the Web many times. The first generation of Silicon Valley giants got their start in a garage, but they took decades to get big. Now companies start in dorm rooms and get big before their founders can graduate. You know why. Computers amplify human potential: they not only give people the power to create but can also spread their ideas quickly, creating communities, markets, even movements.

Now the same is happening with physical stuff. Despite our fascination with screens, we still live in the real world. It's the food we eat, our homes, the clothes we wear, and the cars we drive. Our cities and gardens; our offices and our backyards. That's all atoms, not bits.

This construction—"atoms" versus "bits"—originated with the work of a number of thinkers from the MIT Media Lab, starting with its founder, Nicholas Negroponte, and today most prominently exemplified by Neal Gershenfeld and the MIT Center for Bits and Atoms. It is shorthand for the distinction between software and hardware, or information technology and Everything Else. Today the two are increasingly blurring as more everyday objects contain electronics and are connected to other objects, the so-called Internet of Things. That's part of what we'll be talking about here. But even more, we'll look at how it's changing manufacturing, otherwise known as the flippin' Engine of the World Economy.

The idea of a "factory" is, in a word, changing. Just as the Web democratized innovation in bits, a new class of "rapid prototyping" technologies, from 3-D printers to laser cutters, is democratizing innovation in atoms. You think the last two decades were amazing? Just wait.

If Fred Hauser were born in 1998, not 1898, he'd still have his workshop, tinkering with nature and bountiful ideas. The only thing that would have changed in his converted garage is the addition of a computer and an Internet connection. But what a change!

Rather than a solo obsession, he likely would have been part of a community of equally obsessed people from around the world. Rather than inventing everything from scratch, he would have built on the work of others, compressing decades of work into months. Rather than patenting, he might have published his designs online, like other members of his community.

When it came time to make more than a handful of his designs, Hauser wouldn't have begged some manufacturer to license his ideas, he would have done it himself. He would have uploaded his design files to companies that could make anything from tens to tens of thousands of units for him, even drop-shipping them directly to customers. Because his design files were digital, robotic machine tools could make them, saving 90 percent or more in tooling costs. Rather than searching for distributors, he would have set up his own e-commerce website, and customers would have come to him via Google searches, not salesmen.

In short, he would have been an entrepreneur, not just an inventor. That, in a nutshell, is the theme of this book. The history of the past two decades online is one of an extraordinary explosion of innovation and entrepreneurship. It's now time to apply that to the real world, with far greater consequences.

We need this. America and most of the rest of the West is in the midst of a job crisis. Much of what economic growth the developed world can summon these days comes from improving productivity, which is driven by getting more output per worker. That's great, but the economic consequence is that if you can do the same or more work with fewer employees, you should. Companies tend to rebound after recessions, but this time job creation is not recovering apace. Productivity is climbing, but millions remain unemployed.

Much of the reason for this is that manufacturing, the big employer of the twentieth century (and the path to the middle class for entire generations), is no longer creating net new jobs in the West. Although factory output is still rising in such countries as the United

States and Germany, factory jobs as a percentage of the overall work-force are at all-time lows. This is due partly to automation, and partly to global competition driving out smaller factories.

Automation is here to stay—it's the only way large-scale manu-facturing can work in rich countries (see chapter 9). But what can change is the role of the smaller companies. Just as startups are the driver of innovation in the technology world, and the underground is the driver of new culture, so, too, can the energy and creativity of entrepreneurs and individual innovators reinvent manufacturing, and create jobs along the way.

Small business has always been the biggest source of new jobs in America. But too few of them are innovative and too many are strictly local—dry cleaners, pizza franchises, corner groceries, and the like, all of which are hard to grow. The great opportunity in the new Maker Movement is the ability to be both small *and* global. Both artisanal and innovative. Both high-tech and low-cost. Starting small but getting big. And, most of all, creating the sort of products that the world wants but doesn't know it yet, because those products don't fit neatly into the mass economics of the old model.

As Cory Doctorow imagined it a few years ago in a great sci-fi book also called *Makers*,[3] which was an inspiration for me and count-less others in the movement, "The days of companies with names like 'General Electric' and 'General Mills' and 'General Motors' are over. The money on the table is like krill: a billion little entrepreneurial opportunities that can be discovered and exploited by smart, creative people."

Welcome to the New Industrial Revolution.

Chapter 2

The New Industrial Revolution

**What happens when the Web generation
turns to the real world.**

Here's the history of two decades of innovation in two sentences: The past ten years have been about discovering new ways to create, invent, and work together on the Web. The next ten years will be about applying those lessons to the real world.

This book is about the next ten years.

Wondrous as the Web is, it doesn't compare to the real world. Not in economic size (online commerce is less than 10 percent of all sales), and not in its place in our lives. The digital revolution has been largely limited to screens. We love screens, of course, on our laptops, our TVs, our phones. But we live in homes, drive in cars, and work in offices. We are surrounded by physical goods, most of them products of a manufacturing economy that over the past century has been transformed in all ways but one: unlike the Web, it hasn't been opened to all. Because of the expertise, equipment, and costs of producing things on a large scale, manufacturing has been mostly the provenance of big companies and trained professionals.

That's about to change.

Why? Because making things has gone digital: physical objects now begin as designs on screens, and those designs can be shared online as files. This has been happening over the past few decades in factories and industrial design shops, but now it's happening on consumer desktops and in basements, too. And once an industry goes digital, it changes in profound ways, as we've seen in everything from retail to publishing. The biggest transformation is not in the way things are done, but in *who's doing it*. Once things can be done on regular computers, they can be done by anyone. And that's exactly what we're seeing happen now in manufacturing.

Today, anyone with an invention or good design can upload files to a service to have that product made, in small batches or large, or make it themselves with increasingly powerful digital desktop fabrication tools such as 3-D printers. Would-be entrepreneurs and inventors are no longer at the mercy of large companies to manufacture their ideas.

This appeals to the Web generation in a way that tinkering in the workshops of old did not. At the same time, the digital natives are starting to hunger for life beyond the screen. Making something that starts virtual but quickly becomes tactile and usable in the everyday world is satisfying in a way that pure pixels are not. The quest for "reality" ends up with making real things.

This is not just speculation or wishful thinking—it can already be felt in a movement that's gathering steam at a rate that rivals the First Industrial Revolution and hasn't been seen since, well, the Web itself.

Today there are nearly a thousand "makerspaces"—shared production facilities—around the world, and they're growing at an astounding rate: Shanghai alone is building one hundred of them.[4] Many makerspaces are created by local communities, but they also include a chain of gym-style membership workshops called TechShop, run by a former executive of the Kinko's printing and copying chain and aiming to be as ubiquitous. Meanwhile, consider the rise of Etsy, a Web marketplace for Makers, with nearly a million sellers who sold more than $0.5 billion worth of their products on the site in 2011.[5] Or the 100,000 people who come to the Maker Faire in San Mateo each year[6] to share their work and learn from other Makers, just as they do at the scores of other Maker Faires around the world.

Recognizing the power of this movement, in early 2012 the Obama administration launched a program[7] to bring makerspaces into one thousand American schools over the next four years, complete with digital fabrication tools such as 3-D printers and laser cutters. In a sense, this is the return of the school workshop class, but now upgraded for the Web Age. And this time it's not designed to train workers for low-end blue-collar jobs, but rather it's funded by the government's advanced manufacturing initiative aimed at creating a new generation of systems designers and production innovators.

Meanwhile, the rise of "open hardware," another part of what's known as the Maker Movement, is now doing for physical goods what open source did for software. Just as online communities of programmers created everything from the Linux operating system that runs most of today's websites to the Firefox Web browser, new communities of Makers are doing the same with electronics, scientific instrumentation, architecture, and even agricultural tools. There are now scores of multimillion-dollar open-hardware companies (including my own company, 3D Robotics[8]); some of them, such as the Arduino electronics development board, have sold more than a million units. Google, too, has joined the movement, releasing open-hardware electronics to connect to the hundreds of millions of phones and other devices that now run its Android mobile operating system.

What started as a cultural shift—a fascination with new digital prototyping tools and a desire to extend the online phenomenon into real-world impact—is now starting to become an economic shift, too. The Maker Movement is beginning to change the face of industry, as entrepreneurial instincts kick in and hobbies become small companies.

Thousands of Maker projects have raised money on "crowdfunding" sites such as Kickstarter, where in 2011 alone nearly 12,000 successful projects (from design and technology to the arts) raised nearly $100 million[9] (in 2012, that is on track to reach $300 million[10]). Venture capitalists joined in, investing $10 million each into Kickstarter, MakerBot, an open-hardware company making 3-D printers, and Shapeways, a 3-D printing service in 2011, as well as $23 million into Quirky, another Maker marketplace.[11]

Some of the biggest companies in the world of professional product design and engineering are now shifting their focus to the emerging Maker market. Industrial giants such as Autodesk, PTC, and 3D Systems have released free design software for amateurs and even kids, along with service bureaus that let them upload their designs and have them 3-D printed or laser-cut. Like IBM a generation ago, which went from corporate mainframes to personal computers, they are recognizing that their futures lie with regular folks. They are pivoting from professionals to everyone.

In short, the Maker Movement has arrived.

This nascent movement is less than seven years old, but it's already accelerating as fast as the early days of the PC, where the garage tinkerers who were part of the Homebrew Computer Club in 1975 created the Apple II, the first consumer desktop computer, which led to desktop computing and the explosion of a new industry.

Similarly, you can mark the beginnings of the Maker Movement with such signs as the 2005 launch of *Make* magazine, from O'Reilly, a legendary publisher of geek bibles, and the first Maker Faire gatherings in Silicon Valley. Another key milestone arrived with RepRap, the first open-source desktop 3-D printer, which was launched in 2007. That led to the MakerBot, a consumer-friendly 3-D printer that is inspiring a generation of Makers with a mind-blowing glimpse of the future of desktop manufacturing, just as the first personal computers did thirty years before.

Makers united

What exactly defines the Maker Movement? It's a broad description that encompasses a wide variety of activities, from traditional crafting to high-tech electronics, many of which have been around for ages. But Makers, at least those in this book, are doing something new. First, they're using digital tools, designing onscreen, and increasingly outputting to desktop fabrication machines. Second, they're the Web generation, so they instinctively share their creations online. By simply bringing the Web's culture and collaboration to the process of making, they're combining to build something on a scale we've never seen from DIY before.

What the Web taught us was the power of "network effects": when you connect people and ideas, they grow. It's a virtual circle—more people combined create more value, which in turn attracts even more people, and so on. That's what has driven the ascent of Facebook, Twitter, and practically every other successful company online today. What Makers are doing is taking the DIY movement online—"making in public"—which introduces network effects on a massive scale.

In short, the Maker Movement shares three characteristics, all of which, I'd argue, are transformative:

1. People using digital desktop tools to create designs for new products and prototype them ("digital DIY").

2. A cultural norm to share those designs and collaborate with others in online communities.

3. The use of common design file standards that allow anyone, if they desire, to send their designs to commercial manufacturing services to be produced in any number, just as easily as they can fabricate them on their desktop. This radically foreshortens the path from idea to entrepreneurship, just as the Web did in software, information, and content.

Nations have always had their tinkerers and inventors. But the shift to digital changes everything about the ability to get those ideas and inventions produced and sold. Workshops of the world, unite!

Today the Maker Movement is where the personal computer revolution was in 1985—a garage phenomenon bringing a bottom-up challenge to the ruling order of the time. As then, the sudden liberation of industrial technology inspires exuberant imagination and some sweeping predictions (including here). The leaders of the Maker Movement echo the fervor of Steve Jobs, who saw in the personal computer not just the opportunity to start a company but also a force that would change the world.

But don't forget: he was right.

Indeed, Jobs himself was inspired by his Maker upbringing. Writing in *Wired*,[12] Steven Levy explained the connection, which led to the original Apple II in 1977:

> His dad, Paul—a machinist who had never completed high school—had set aside a section of his workbench for Steve, and taught him how to build things, disassemble them, and put them together. From neighbors who worked in the electronics firm in the Valley, he learned about that field—and also understood that things like television sets were not magical things that just showed up in one's house, but designed objects that human beings had painstakingly created. "It gave a tremendous sense of self-confidence, that through exploration and learning one could understand seemingly very complex things in one's environment," he told [an] interviewer.

Later, when Jobs and his Apple cofounder, Steve Wozniak, were members of the Homebrew Computer Club, they saw the potential of desktop tools—in this case the personal computer—to change not just people's lives, but also the world.

In this, they were inspired by Stewart Brand, who had emerged from the psychedelic culture of the 1960s to work with the early Silicon Valley visionaries to promote technology as a form of "computer liberation," which would free both the minds and the talents of people in a way that drugs had not.

In his biography of Steve Jobs, Walter Isaacson describes Brand's role in the origins of what is today the Maker Movement:

Brand ran the Whole Earth Truck Store, which began as a roving truck that sold useful tools and educational materials, and in 1968 he decided to extend its reach with *The Whole Earth Catalog*. On its first cover was the famous picture of Earth taken from space; its subtitle was "Access to Tools." The underlying philosophy was that technology could be our friend. Brand wrote on the first page of the first edition, "A realm of intimate, personal power is developing—power of the individual to conduct his own education, find his own inspiration, shape his own environment, and share his adventure with whoever is interested. Tools that aid this process are sought and promoted by *The Whole Earth Catalog*." Buckminster Fuller followed with a poem that began, "I see God in the instruments and mechanisms that work reliably."[13]

The Homebrew Computer Club, where Jobs and Wozniak brainstormed the first Apple computer, was founded on these principles. Today it carries on in hundreds of makerspaces, each using twenty-first-century tools to try to effect the same sort of revolutionary social and economic change.

Real countries make stuff

Any country, if it wants to stay strong, must have a manufacturing base. Even today, about a quarter of the U.S. economy consists of the manufacturing of physical goods. When you include their distribution and sale in retail outlets, you're talking about closer to three-quarters of the economy. A service economy is all well and good, but eliminate manufacturing and you're a nation of bankers, burger flippers, and tour guides. Software and information industries get all the press, but they employ just a small percentage of the population.

Some of us say that we "live online," but it's not true when it comes to spending or living our everyday lives. Our commercial lives reside mostly in the real world of bricks and mortar, of food and clothes, of cars and houses, and, until some sci-fi future arrives where we're just disembodied brains in vats, that will continue to be the case. Bits are thrilling, but when it comes to the overall economy, it's all about atoms.

Yet the cost of labor has made it harder and harder to keep manufacturing industries going in the rich countries of the West. Driven by the exodus of factory jobs due largely to Asian cost advantages, manufacturing employment in the United States is at a century-long low, both in absolute numbers and as a percentage of total working population. What's worse, those factories that are bucking the trend are having trouble finding qualified workers, as a generation has turned away from manufacturing as a career option. The industry that created the middle class in America is now seen to be in terminal decline (as we'll see later, this isn't the case, but without a reset, appearances risk becoming reality). Working in a factory sounds boring, dangerous, and dead-end.

But today we have a path to reverse that—not by returning to the giant factories of old, with their armies of employees, but by creating a new kind of manufacturing economy, one shaped more like the Web itself: bottom-up, broadly distributed, and highly entrepreneurial.

It is almost a cliché that anyone with a sufficiently good software idea can create a fabulously successful company on the Web. That's because there are practically no barriers preventing entry to entrepreneurship online: if you've got a laptop and a credit card, you're in business.

But manufacturing was always seen as something else entirely. Making stuff is expensive; it needs equipment and skills in everything from machining to supply-chain management. It usually requires huge up-front investments, and mistakes lead to warehouses of unsellable inventory. Failure may be celebrated online, where the cost of entry is relatively low, but in the world of making stuff, failing means ruination. Atoms are weighty, and so are the consequences of their failure. When you shut down a website, nobody cares. When you shut down a factory, lots of people lose their jobs, and the debts can haunt the owners for the rest of their lives.

Or at least that's the way it used to be. But over the past few years, something remarkable has happened. The process of making physical stuff has started to look more like the process of making digital stuff. The image of a few smart people changing the world with little more than an Internet connection and an idea increasingly describes manufacturing, too.

DIY manufacturing

Why? Because even commercial manufacturing itself has become digital, networked, and increasingly open—just like the Web. The biggest manufacturing lines speak the same language as a Maker-Bot ("G-code"), and anyone can move from one to the other. As a result, global manufacturing can now work at any scale, from units of one to millions. Customization and small batches are no longer impossible—in fact, they're the future.

It's like the photo management software, such as Picasa or iPhoto, that you probably already use on your own computer. They have a menu that allows you to choose whether to print your photos on your desktop printer or upload them to a service bureau to be professionally printed, or even bound into a photo album. The same ability has come to desktop CAD tools, where you can design 3-D objects onscreen. Once you've created something in a CAD program, you can choose whether to "print local" (prototype one copy on your 3-D printer or other desktop fabricator) or "print global" (send it off to a service bureau to be manufactured in volume). The only real difference is that sending it off to a service bureau adds a credit-card or invoice step, just like the photo printing services you already use.

This ability—to manufacture "local or global" at will—is a huge advantage. That simple menu option compresses three centuries of industrial revolution into a single mouse click. If Karl Marx were here today, his jaw would be on the floor. Talk about "controlling the tools of production": you (you!) can now set factories into motion with a mouse click. The distinction between amateur and entrepreneur has been reduced to a software option. The step from making one to making thousands is simply a matter of what menu options you select and how much you want to pay (or put on your credit card).

You can already see this in Autodesk's free 123D CAD program, which has a "Make" menu option that walks you through the choice between desktop prototyping and service bureaus. Over time, more such CAD programs will come with software "wizards" that can help you choose whether to fabricate in 2-D or 3-D, choose different materials based on their physical properties and costs, and integrate off-the-shelf parts that the service bureau can order for you. Companies such as Ponoko already provide this sort of online service, serving as the Web link that connects desktop tools to global manufacturing capacity, which will eventually power the "Make" button in the program you use to create anything. The expertise of the machine shop is being replicated in software algorithms.

The reinvention of the sprinkler

Remember my grandfather's automatic sprinkler and my thought experiments in how differently its creation would have played out if he had invented it today? Rather than having to patent it and license it to a manufacturer (and lose control of his invention in the process), he could have brought it into production himself, becoming not just an inventor but also an entrepreneur.

Well, rather than just imagining what that would have been like, I thought it would be interesting to try it. So I decided to reinvent the automatic sprinkler system in the modern Maker model.

I am, it must be said, not a natural sprinkler entrepreneur. For starters, our "lawn" is ten feet long and four feet wide (the perils of living in the Berkeley hills); you can mow it with a pair of shears. I have absolutely no interest in gardening, and set foot on the grass only about once a year to set up a pup tent so the kids can conduct their annual adventure in "camping." My wife is the gardener, and she guards the flowerbeds with an iron fist; she was clear from the start that we would be doing no sprinkler experimentation in her domain.

But because my grandfather's big idea was the automatic sprinkler, for the sake of the family legacy a sprinkler it must be. So I talked to friends with proper lawns and sprinkler systems, visited garden stores, and started reading gardening sites. If I were to become a sprinkler inventor *and* entrepreneur, what problems would I be solving?

My assumption was that the best way to reinvent a mature industry would be to open it up to the ideas of others. So I asked a few basic questions, which you could call a toolkit for transformation (it can apply to practically any product):

1. How would these products be improved if they were connected to the Internet?

2. How would they be improved if the designs were open, so anyone could modify or improve them?

3. How much cheaper would they be if their manufacturers didn't charge for their intellectual property?

It didn't take me long to decide that sprinklers, despite my grandfather's wisdom and the collective innovation of a huge industry built over half a century or more, could be made a lot better. For starters, all the products on the market were proprietary, which meant that even if they did connect to the Internet (and few did), you had to pay a service fee for the privilege and were limited to what the manufacturer allowed. You could connect only the sensors that the manufacturer sold, and use them only the way the manufacturer had provided for. And they were expensive: a full installation could easily run into the thousands of dollars and typically needed a consultant.

Now imagine a way better sprinkler—call it OpenSprinkler.

First, let's make it easy to control the sprinkler with your phone. Left for a vacation but forgot to set the sprinkler system? There's an app for that. Want to know what the soil moisture level is in the strawberry patch on a hot day while you're at work? Just check your pocket.

What if your sprinkler system knew it was going to rain tomorrow, so it didn't have to water today? Sure, you can buy high-end proprietary systems that will do that, but you have to pay a subscription fee. And if you have a better local weather data source than the one they use, you're out of luck—you are stuck with theirs. Let's make that free and open, too.

What if you don't want to have to read the manual just to figure out how to use your sprinkler system's cryptic menus? With OpenSprinkler you can set it up on a simple website with an easy-to-use graphic interface. And if you don't like the control panel we created, there are dozens of others to choose from, thanks to a community encouraged to create their own.

So there you have it, a recipe for a better sprinkler: open, Internet-connected, and inexpensive.

Easy enough to imagine. But how to make it real?

My electronics company, 3D Robotics, is based on an open-source computing platform called Arduino, which is a cheap and easy-to-use processor and free programming environment. It allows anyone to connect computing and the physical world, by making it easy to attach sensors and actuators to a computer program. This is often called "physical computing" or "embedded computing," and you see examples of it all around you. Practically every electronic device in your home works this way, from your thermostat to your alarm clocks, stereos, microwave oven, and portable music players. Your car has dozens of embedded computers. The difference is that they are all closed and proprietary, while Arduino is designed to be easy for anyone to use and modify. Much of the emerging "Internet of Things" movement is built on Arduino-based devices connected to the Web, from cof-feemakers that tweet their status to pet feeders you can control from your phone, wherever you are.

So, because I knew it best, I decided to base the sprinkler control-ler on Arduino. That meant it could tap into a huge community of people who are using Arduino for all sorts of other purposes, and who had already solved most of the problems of connecting it to the Internet and any sensor you can imagine. My hope was that by using Arduino, most of my work would already have been done.

A quick search confirmed that this was the case; indeed, it showed that there was already a quite active Arduino sprinkler subculture. There were countless projects to control drip irrigation, to monitor soil moisture, even to steer plant containers toward the sun. Why so many? Well, most of it was simply putting together two geeky passions—gardening and computing—but the truth is that some were also driven by hydroponic "gardeners," who I assume are mostly peo-ple growing high-quality pot. Now there's a market not well served by the traditional sprinkler makers!

Nevertheless, there were still improvements to be made, and I found a few like-minded souls: Rui Wang, a University of Massa-chusetts professor who had figured out how to connect Arduino to a cheap commercial water valve that was easily available. And Andrew Frueh, who had started the sophisticated GardenBot project. All they needed was a better way to hook all this computer-controlled garden technology to the Internet, and we'd be in business. A few months of tinkering and we had a very functional prototype. It connected to the Web and thus any weather service online, and had a nifty wireless connection from your home network to a sprinkler controller box that could manage any number of valve networks and sensors.

At that point we had completed the invention stage, which was pretty much as far as my grandfather got on his own. But what would happen next is what shows the difference between then and now. My grandfather was forced to patent his invention, which was an expensive and time-consuming process involving lawyers and piles of paperwork. We, in contrast, just published everything online under open-source licenses. My grandfather had to find a manufacturer who would license his patents and put the sprinkler into production on its own terms. We just had to send the electronics designs to an assembly house (I chose Advanced Circuits, with which I had worked before) and sent the CAD design of the enclosure to a service that would turn it into a mold for injection-molding, which could then be sent to an injection-molding plant that would work at a small scale.

We calculated that an OpenSprinkler controller box, which is to say a Web-connected, easily programmable, cell-phone-friendly sprinkler brain, could be made and sold at a modest profit for about $100. That's between one-third and one-fifth of the price of commercial sprinkler systems with similar features. When your R&D is free (thanks, open-source community!) and you don't charge for intellectual property, it's not hard to undercut proprietary alternatives, even at lower volume.

In fact, it was even cheaper—today you can buy an OpenSprinkler kit for $79.95. Rui Wang used commercial suppliers to make the electronics boards and supply the necessary components, and he set up a Web store to sell it. It cost less than $5,000 to get to market, all told. While that's not pocket change, it's a lot less than my grandfather had to pay just for his patent attorneys' fees. The company that eventually licensed his patent no doubt spent a hundred times that to get a product out the door.

The point is that as entrepreneurship goes, this is dirt cheap. It's within the bounds of a credit-card limit and a tiny fraction of what starting a manufacturing operation used to cost.

One way or another, the sprinkler industry will change over the next few years as other newcomers build projects on Internet-centric, open-innovation models and enter the market. Maybe they will use our work, or maybe they'll come up with better designs of their own. But the point is that the real innovators probably won't be established players in the garden equipment market. Instead, they'll be startups cast more from the Web model. Today entrepreneurship is a choice in the way it never was for my grandfather.

And now for everything else

If sprinklers aren't your thing, you can substitute almost any other product or industry. Just in the past half hour as I was writing this, my news feeds brought me reports of similar Web-enabled hardware projects in horse management (electronics in barns that track animal comings and goings; apparently that's something ranchers need), home thermostats, biology lab centrifuges, and weather stations. Organizations as large as the Pentagon's research group—the Defense Advanced Research Program Administration (DARPA)—and General Electric are using open innovation for creating everything from small drones for the Army to smart electric outlets in your home.

Of course the New Industrial Revolution is not limited to open innovation. Conventional proprietary product development benefits from the same desktop prototyping tools, from 3-D printers to CNC (computer numerical control) routers. These new capabilities are accelerating innovation in the biggest companies in the world, from Ford's automobile interiors to IKEA's new kitchenware. As we'll see later, companies such as General Electric are using Maker-like community innovation methods among their own employees to develop proprietary products—open innovation doesn't have to be *wide* open. Midsized manufacturing companies in the United States and Europe are increasingly able to compete with low-cost labor in China by using digital manufacturing techniques to automate what used to require either lots of human labor or ruinously expensive equipment and tooling.

Behind all of them are the same thing: people working together with extraordinary new tools to create a manufacturing revolution. The shape of the twenty-first century's industrial structure will be very different from the twentieth century's. Rather than top-down innovation by some of the biggest companies in the world, we're seeing bottom-up innovation by countless individuals, including amateurs, entrepreneurs, and professionals. We've already seen it work before in bits, from the original PC hobbyists to the Web's citizen army. Now the conditions have arrived for it to work again, at even greater, broader scale, in atoms.

Chapter 3

The History of the Future

**What happened in Manchester and the cottage
industries of England changed the world.
It could happen again.**

In 1766, James Hargreaves, a weaver in Lancashire, was
visiting a friend when he saw a spinning wheel fall on its side. For
some reason it kept spinning, and something about the contrap-
tion still working in the unfamiliar orientation triggered a vision in
Hargreaves's mind: a line of spindles, side by side, spinning multiple
threads of cotton from flax simultaneously. When he returned home,
he started whittling up just such a machine from spare wood, with
the spindles connected by a series of belts and pulleys. Many versions
later, he had invented the spinning jenny, a pedal-powered device that
could allow a single operator to spin eight threads at the same time
(*jenny* was Lancashire slang for "machine").

The machine amplified the output of a single worker by a factor of
eight at the start, and could easily be expanded beyond that. And this
was just the beginning.

There was nothing new about textile-making machines themselves. The ancient Egyptians had looms, after all, and the Chinese
had silk-spinning frames as early as 1000 BCE. The hand-powered
spinning wheel was introduced in China and the Islamic world in
the eleventh century, and the foot treadle appeared in the 1500s. You
only have to look at illustrated fairy tales to see spinning wheels in
widespread use.

But the earlier machines didn't launch an industrial revolution, while Hargreaves's invention, along with the steam engine and even more sophisticated power looms that came later, did. Why? Historians have been debating this for centuries, but they agree on a few reasons. First, unlike silk, wool, and hemp, which were used in many of the earlier machines, cotton was a commodity that could reach everyone. It was simply the cheapest and most available fiber in the world, even more so once the expanding British trade empire brought bales of the stuff from India, Egypt, and the New World.

Second, the spinning jenny, being driven by a series of belts and pulleys, was designed to distribute power from a central point to any number of mechanisms operating in parallel. Initially that was human muscle power, but the same principle could use much stronger motive forces—first water, then steam—to drive even more spindles. In other words, it was a scalable mechanism, able to take advantage of bigger sources of power than just arms and legs.

Finally, it arrived at the right time, in the right place. Britain in the 1700s was going through an intellectual renaissance, with a series of patent laws and policies that gave artisans the incentive not only to invent but also to share their inventions.

As William Rosen put it in his 2010 book, *The Most Powerful Idea in the World*:

Britain's insistence that ideas were a kind of property was as consequential as any idea in history. For while the laws of nature place severe limits on the total amount of gold, or land, or any other traditional form of property, there are (as it turned out) no constraints at all on the number of potentially valuable ideas. . . . The Industrial Revolution was, first and foremost, a revolution in *invention*. And not simply a huge increase in the number of new inventions but a radical transformation in the process of invention itself.[14]

In June 1770, Hargreaves submitted a patent application, number 962, for a version of the spinning jenny that could spin, draw, and twist sixteen threads simultaneously. The delay between this patent application and his first prototypes meant that others were already using the jenny by the time his patent was granted, making it difficult for him to enforce his patent rights. Even worse, the machine made enemies.

Starting in Hargreaves's native Lancashire, the spinning jenny's magical multiplication of productivity was initially, as you might expect, little welcomed by the local artisans, whose guilds had controlled production for centuries—they hated it. As yarn prices started to fall and opposition from local spinners grew, one mob came to his house and burned the frames for twenty new machines. Hargreaves left for Nottingham, where the booming cotton hosiery industry needed more cotton thread. He died a few years later, in 1778, having made a little money from his invention, but still far from rich.

While this was happening, the American colonies were declaring independence and war. James Watt invented the steam engine in 1776. Although the exact timing with the Declaration of Independence is a coincidence, the connection between the two is not. Britain was finding it increasingly difficult to support its empire on resource extraction from its colonies alone, especially as they became more difficult to manage. It needed to increase production at home, where the political and military costs were lower. Mechanized planting and harvesting tools were already hugely increasing the output of British farms. The arrival of machines to turn agricultural commodities into goods that could be sold around the world promised the opportunity to shift from a nation that commanded global power by force to one that used trade instead. But its greatest impact was initially at home, where the immediate effect was to both reshape the landscape and hugely elevate the living standard of millions of Britons.

What revolutions can do

What exactly is an "industrial revolution"? Historians have been debating this since the late eighteenth century, when they first noticed that something startling was happening to growth rates. It was already obvious that the manufacturing and trade boom that came with the first factories had changed the economy, but the sheer magnitude of it wasn't immediately clear, in part because statistics were hard to find. But by the 1790s, the effects didn't need an accountant to observe them. Populations were simply exploding, and for the first time in history, wealth was spreading beyond landed gentry, royalty, and other elites.

Between 1700 and 1850, the population of Great Britain tripled. And between 1800 and 2000, average per capita income, inflation adjusted, grew *tenfold*. Nothing like this had ever happened before in recorded history. It seemed clear that this social revolution was connected somehow to the industrial quarters that were increasingly dominating England's fast-growing cities. But why mechanization led to population growth, to say nothing of the other booming quality of life measures, took longer to figure out.

There was, of course, more to it than just factories. Improved farming methods, including the fencing in of pastures that avoided the "tragedy of the commons" problem, had a lot to do with it. And more children were living to adulthood, thanks to the invention of the smallpox vaccine and other medical advances. But industrialization helped even more.

Although we think of factories as the "dark satanic mills" of William Blake's phrase, poisoning their workers and the land, the main effect of industrialization was to improve health. As people moved from rural communities to industrial towns, they moved from mud-walled cottages to brick buildings, which protected them from the damp and disease. Mass-produced cheap cotton clothing and good-quality soap allowed even the poorest families to have clean clothing and practice better hygiene, since cotton was easier to wash and dry than wool. Add to that the increased income that allowed a richer and more varied diet and the improved access to doctors, schools, and other shared resources that came with the migration to cities, and whatever ill effects resulted from working in the factories were more than compensated for by the positive effects of living around them. (To be clear, working in factories was tough, with long hours and poor conditions. But the statistics suggest that working on farms was even worse.)

The difference between life before and after this period is really quite amazing. Our modern expectation of continual growth and improving quality of life is just a few hundred years old. Before that, things stayed more or less the same, which is to say pretty bad, for thousands of years. Between 1200 and 1600, the average life span of a British noble (for whom records were best kept) didn't go up by so much as a single year.[15] Yet between 1800 and today, life expectancy for white males in the West doubled, from thirty-eight years to seventy-six. The main difference was the decline in child mortality. But even for those who survived childhood, life expectancy grew by about twenty years over that period, a jump of a magnitude never before seen.

The explanation for this had to do with all sorts of changes, from improvements in hygiene and medical care to urbanization and education. But the common factor is that as people got richer, they got healthier. And they got richer because their abilities were being amplified by machines, in particular machines that made stuff. Of course, humans have been using tools since prehistory and one could argue that the "technologies" of fire, the plow, domesticated animals, and selective breeding were as defining as any steam engine. But agricultural technologies just allowed us to feed more people more easily. There was something different about the machines that allowed us to make products that improved our quality of life, from clothes to transportation.

For one thing, people around the world wanted such goods, so they drove trade. Trade, in turn, drove the engine of comparative advantage, so that countries did what they could do best and imported the rest, which improved everyone's productivity. And that, in turn, drove growth. As went the cotton mills of Manchester, so went the world economy.

The Second Industrial Revolution

The term *industrial revolution* itself was coined in 1799 by Louis-Guillaume Otto, a French diplomat, in a letter reporting that such a thing was under way in France (revolutions were much in vogue).[16] *Revolution* was also, perhaps unsurprisingly, the term used to describe the industrial changes by Friedrich Engels, whose capitalist critiques in the mid-1800s helped lead to Marxism. And it was popularized in the late 1800s by Arnold Toynbee, a British economic historian who gave a series of famous lectures on why this industrial movement had had such a profound impact on the world economy.

But at its core, *industrial revolution* refers to a set of technologies that dramatically amplify the productivity of people, changing everything from longevity and quality of life to where people live and how many there are of them.

For example, around 1850, the rise of the factory (from *manufactory*, as it was originally known) was joined by another technological wave, the development of steam-powered ships and railroads, which brought similar productivity gains to transportation. The invention of the Bessemer process for making steel in large quantities in the 1860s led to mass production of metal goods and eventually the assembly line.

Combined with the rise of the chemical industries, petroleum refining, and the internal combustion engine and electrification, this next phase of manufacturing transformation is called by many historians the "Second Industrial Revolution." They place it from 1850 to around the end of World War I, which includes Henry Ford's Model-T assembly line, with its innovations of stockpiles of interchangeable parts and the use of conveyer belts, where products being produced moved to stationary workers (who each did a single task), rather than the other way around.

Today, in a fully industrialized economy, we forget just how much the First and Second Industrial Revolutions changed society. We talk in terms of productivity enhancements, but consider what that means in terms of people's lives. When we moved from hunter-gatherers to farmers, one person could feed many. We were able to break out of the cycle of most other animals, where everyone's job is to feed themselves or their offspring, and pursue division of labor, where we each do what we do best. This created spare time and energy, which could be invested in such things as building towns, inventing money, learning to read and write, and so on.

What the spinning jenny and its kin had created was an inflection point in the arc of history, a radical shift in the economic status quo. It elevated our species to one that was less about what we could do and more about what we knew. We became more valuable for our brains than for our muscles. And in the process, it made us richer, healthier, longer-living, and hugely more populous. Revolutions should be measured by their impact on people's lives, and as such the First Industrial Revolution is unparalleled.

The move from hand labor to machine labor freed up people to do something else. Fewer people in society were needed to create the bare essentials of food, clothing, and shelter, so more people could start working on the nonessentials that increasingly define our culture: ideas, invention, learning, politics, the arts, and creativity. Thus the modern age.

Writer Vankatesh Rao argues that the main effect of this was on time. Machines allow us to work faster, doing more in less time. That liberates those hours for other activities, whether productive or leisure. What the First Industrial Revolution did create, more than anything else, was a vast surplus of time, which was reallocated to invent practically everything that defines the modern world. Four hundred years ago, nearly everyone you'd know would be involved in producing the staples of existence: food, clothing, shelter. Today, odds are, almost none of them are. Rao writes:

> The primary effect of steam was not that it helped colonize a new land, but that it started the colonization of time. Many people misunderstood the fundamental nature of Schumpeterian growth [a reference to the innovation and entrepreneurship growth theories of the economist Joseph Schumpeter] as being fueled by ideas rather than time. Ideas fueled by energy can free up time which can then partly be used to create more ideas to free up more time. It is a positive feedback cycle.[17]

The Third Industrial Revolution?

There are those who argue that the Information Age is the Third Industrial Revolution. Computing and communications are also "force multipliers," doing for services what automation did for manufacturing. Rather than amplifying human muscle power, they amplify brain power. They can also drive productivity gains in existing industries and create new ones. And by allowing us to do existing jobs faster, they free us up to do new ones.

But in the same way that the first two Industrial Revolutions required a series of technologies to come together over many decades before their true impact was felt, the invention of digital computing is not enough by itself. The first commercial mainframes replaced some corporate and government accounting and statistics jobs; the first IBM PCs replaced some secretarial jobs. Neither changed the world.

Only when the computers were combined with networks, and ultimately the network-of-all-networks, the Internet, did they really start to transform our culture. And even then the ultimate economic impact of computing may not be felt mostly in the services transformed by software (although there are a lot of them), but rather by how they transform the same domain as the first two Industrial Revolutions: the work of making stuff itself.

In short, the dawn of the Information Age, starting around 1950 and going through the personal computer in the late 1970s and early 1980s and then the Internet and the Web in the 1990s, was certainly a revolution. But it was not an *industrial* revolution until it had a similar democratizing and amplifying effect on *manufacturing*, something that's only happening now. Thus, the Third Industrial Revolution is best seen as the combination of digital manufacturing and personal manufacturing: the industrialization of the Maker Movement.

The digital transformation of making stuff is doing more than simply making existing manufacturing more efficient. It's also extending manufacturing to a hugely expanded population of producers—the existing manufacturers plus a lot of regular folk who are becoming entrepreneurs.

Sound familiar? It's exactly what happened with the Web, which was colonized first by technology and media companies, which used it to do better what they already did. Then software and hardware advances made the Web easier to use for regular folks (it was "democratized"), and they charged in with their own ideas, expertise, and energy. Today the vast majority of the Web is built by amateurs, semipros, and people who don't work for big technology and media companies.

We talk a lot about the "weightless economy," the trade in intangible information, services, and intellectual property rather than physical goods (the weightless economy consists of anything that doesn't hurt your foot if dropped upon it). Yet as big as the economy of bits may be, that dematerialized world of information trade is a small fraction of the manufacturing economy. So anything that can transform the process of making stuff has tremendous leverage in moving the global economy. That's the making of a real revolution.

Let's return to Manchester to consider how that might work in the real world.

Manchester, yesterday and tomorrow

Manchester is a city defined by its rapid rise long ago, and an agonizingly slow fall ever since. Today, in its manufacturing museum and crumbling warehouse districts, we see mostly the lost past: nostalgia for a time when Manchester was the world's greatest industrial city and the skyline was punctuated with the smokestacks of the world's clothes-makers. Every great city has its defining moment, and Manchester's can be seen in the architecture of the semirenovated Northern Quarter, which is still dominated by colossal Victorian brick warehouses and former factory buildings.

Why did the First Industrial Revolution take off in Manchester? There were other cities and regions that had early factories, including Birmingham and smaller towns in Lancashire. But Manchester had several key advantages. First, it had plenty of free space and relaxed building laws, which made it possible to build factories and housing for the workers, something that would have been hard in the more built-up and restrictive port cities such as Liverpool. It was near rivers and streams that could provide waterpower for the early mill-driven factories. The largest of those rivers, the Mersey, extended all the way to the Atlantic, making it relatively easy to bring raw materials in and send finished goods out. And it was eventually well connected with rail lines, which brought coal from elsewhere in England and Wales.

In the mid-1800s, Manchester was at its peak. England grew hardly any cotton, but Manchester was called "Cottonopolis." Bales of raw cotton came in by sea from far-off lands and were transformed by miraculous machines—combing, tight weaving, and precision dyeing—into thread, cloth, and finally clothes. Then those goods were sent off through the same channels to markets around the world. It was a glimpse of the future: global supply chains, competitive advantage, and automation made a once-unremarkable city the center of the global textile trade.

Impressive as the new manufacturing machines were, the supply networks that fed them were equally important. Bigger, more efficient factories needed more and cheaper raw materials—not just cotton from Egypt and the Americas, but dyes and silk from Asia and eventually mineral resources such as iron ore and coal. That's why the steam engine's impact was felt as much in the evolution of sailing ships to steam freighters and the rise of steam locomotives as it was in the factory. Every step in the supply chain had to get more efficient for the impact of mechanized production to be felt.

At their height, Manchester's canals were the communications channels of the First Industrial Revolution. It was not enough to make stuff efficiently; it had to be distributed efficiently, too. Smaller canal projects eventually led to the Manchester Ship Canal in 1884, which allowed oceangoing freighters to sail right up to the Port of Manchester, forty miles inland. It was the perfect combination: an inland city with room for industrial expansion that, thanks to the big canal, could ship goods nearly as efficiently as a port city. Meanwhile, the railroads were doing the same on land: Manchester became one end of one of the world's first intercity rail lines, the Liverpool and Manchester Railway.

As a result, Manchester's manufacturing became the envy of the world, and companies everywhere sought to copy its model. Sadly for the local factories, they could. Along with selling clothing, Manchester firms started selling the machines that made them. Companies such as J&R Shorrocks and Platt Brothers, which were famed for their engineering skills, soon were exporting their machinery around the world, where it was copied, enhanced, and otherwise commoditized. By the 1900s, huge textile factories could be found from France to America. Manchester's mechanical advantages had been matched, and new industrial centers closer to agricultural sources of the raw cotton, especially in the American South, began to take over.

Manchester's factories went through the long-familiar quest to move upstream, with more-fashionable designs, higher quality, branded appeal, and further mechanical innovation. It certainly helped, and averted what might have been an overnight implosion of an industry in the face of cheaper competitors. Instead, Manchester's textile decline stretched out over a century. But by the 1950s, there were more empty factories than full ones, and the city had become a symbol of Britain's lost industrial might.

By the 1980s, the city was better known for the raves held in empty warehouses than for what had once filled them. Not for nothing was the music label that was behind the UK's Manchester-based post-punk scene of the 1980s (Joy Division, New Order, Happy Mondays, and many others) called Factory Records—it started with a series of music clubs housed in former Victorian factories. Manchester had become a symbol of manufacturing decline. Young people with not enough to do created a thriving music scene, but their joblessness and existential despair also spoke to the vacuum left in the birthplace of the First Industrial Revolution.

In 1996, the IRA parked a truck packed with explosives in the city center. Although a warning call ensured that the area was evacuated before the bomb exploded, it badly damaged dozens of buildings. This became something of a turning point for Manchester. After years of decline and failed turnaround strategies, reconstruction became a catalyst. The tragedy focused national attention on the downtrodden city, and provided an opportunity to rethink the city center.

Today, that is well under way. In Manchester's center today is Spinningfields, which in the 1880s was a packed district of textile factory complexes, each employing as many as fifteen thousand women working power looms and sewing machines. Today, Spinningfields is a modern office and shopping district, with high-end boutiques and dramatic architecture. Its industrial past is reflected in the two-story windows of one clothing store, which displays an art installation matrix of hundreds of old Singer sewing machines. The clothes inside are mostly made in China, of course.

A few blocks north of Spinningfields is the Northern Quarter, where some of those original textile warehouses have been gutted and rearchitected as high-design workspaces, which are now filling with Web companies, game developers, and graphics studios. This is the showpiece of Manchester's hoped-for reinvention as a digital hub. Perhaps the design and engineering skills that powered the Industrial Age are still there, ready to be recast in media, entertainment, and marketing. (It's still too early to say; much of the space remains to be filled, and there is a fair amount of government money propping up what's there.)

But walk a few blocks farther north to the optimistically titled New Islington quarter (a reference to a posh district of London), and the reinvention of Manchester is more uncertain. Here lie mostly ruins: Victorian factories that are now empty shells, with caved-in roofs and long-gone windows. They are listed as historic buildings, so they cannot be bulldozed, but the cost and risk of rebuilding them with original façades intact (as the listing requires) as modern buildings are too high. So they are left to decay, reminders of empires lost. A few others did catch the eye of investors during the recent real-estate bubble, but it ended badly. Today they are fenced-in construction sites with very little active construction actually going on, frozen between the past and the future, and in the protracted present they give the area the feel of a massive worksite without workers, all gravel and dust and no life.

Yet amid this postindustrial landscape are pockets of hope and growth. One of them is on a former factory site next to a former cholera hospital, on the banks of one of Manchester's many canals. Here a huge modern building stands, with stacks of floors, each angled a bit from the one below and painted with tastefully matched accent colors of pink, brown, and peach. Called Chips, supposedly because the architect piled up french fries ("chips") to brainstorm its shape, it was designed to be the model of a modern work/live/play space. The upper floors are built as condominiums. The lower floors are designed for restaurants and shops. And in the middle are floors for offices and workspaces.

Needless to say, the bursting of the real-estate bubble, which halted most of the construction in the area, pretty much put a halt to any restaurant and café plans around the building, and not many homeowners wanted to live among worksites. So rather than leave the building empty, the owners decided to try an experiment that evoked Manchester's beginning: they offered it to the regional manufacturing association as the site of a laboratory in the future of making stuff. Today it is the Manchester Fab Lab, the first Fab Lab in the United Kingdom.

Fab Labs are a special kind of makerspace. They are built on a model developed a decade ago by Neil Gershenfeld's Center for Bits and Atoms—the labs grew out of Gershenfeld's popular class at MIT called "How to Make (Almost) Anything." Each Fab Lab (as of this writing there are fifty-three of them, in seventeen countries around the world), has at least a minimal set of digital fabrication tools: a laser cutter, a vinyl cutter, a big CNC machine for furniture and a small one for circuit boards, basic electronics equipment, and sometimes a 3-D printer, too. They sometimes have more traditional machine shop tools such as metal lathes and drill presses, but typically they are focused on smaller-scale prototyping.

Fridays and Saturdays are free to all at the Fab Lab Manchester. On a typical Friday while I was there, there was a gentle hum of activity as students from local universities worked on architecture and furniture models, and the laser cutter was in constant use making art pieces and design-school classwork. Projects made on free days are supposed to be documented online so others can share them. On other days, members pay to use the facility, and those projects can be proprietary and closed.

It is, to be honest, a little hard to see this makerspace as the seed of a new British manufacturing industry. Most of the work is being done by local students, and is the sort of modest stuff you might expect to find in any design or shop class. No hot startups have been spawned here yet; unlike such makerspaces as TechShop in the United States, the place is not abuzz with entrepreneurship. But Haydn Insley, the lab manager, sees the experiment as more about liberating creativity. "It's about the ability for individuals to make—and, more importantly, modify—anything. Everyone here has an idea—we're trying to make it easier for them to realize it. What becomes important is the designs, not the fabrication."

When you look at the UK manufacturing success stories that still exist today, you can see where Insley gets his optimism. Although textiles and flatware are long gone, the UK still has a major aerospace industry (British Aerospace, or BAE Systems as it is now called, is the world's second-largest defense contractor), and its car designs are still world renowned. And then there are innovative consumer product companies such as Dyson, which uses high design and superior engineering to get consumers to pay premium prices in previously stale and commoditized market segments such as vacuums and fans. Manchester's universities still produce more engineers than universities in any other city in the UK. The skills are there—they're just looking for new outlets.

Maybe one of the dreadlocked design students hovering over the laser cutter in the Manchester Fab Lab will be the next Dyson. Or maybe they're working on their own, using many of the same tools, now cheap enough for an individual to own. The Fab Lab has already created hundreds of projects, and it's just getting started. But here's what we do know: Manchester once made things that changed the world. It's in the water, in the air, woven into the fabric of its history. Whether it will happen again at the Fab Lab, it's now possible to dream of that again. The machines are running again on the Mersey.

But there are some significant differences between then and now. Whereas the First Industrial Revolution could have taken off only in a place like Manchester, with its natural resources and transport infrastructure, this new Maker Movement can occur anywhere. In part for historical resonance, the Manchester Fab Lab is located among the shells of old textile factories, but the tools and technologies within its walls could just as easily be in the offices of a London skyscraper or a converted barn in the countryside. Meanwhile, the Makers using them could be even more widely scattered, uploading design files from their homes. "Place" matters less and less in manufacturing these days—ideas trump geography.

What's more, you don't need a huge factory at all anymore—the days of belching smoke and steel pistons the size of boxcars are gone. Small-scale enterprises can thrive in the new world of distributed manufacturing. Ironically, this is almost a return to the very earliest days of the First Industrial Revolution. The spinning jenny changed the world not by creating the manufacturing plant, but by creating the cottage industry. And the cottage industry can be a very powerful economic force indeed.

What we now know as cottage industries (originally known as "the domestic system" or "outwork system") began with wooden-framed machines with foot pedals that could make many threads at the same time, essentially acting like many spinning wheels operating simultaneously. They were relatively easy to build or cheap to buy, and could be operated in a table-sized space. In a sense, they were the "desktop manufacturing" of the day.

The spinning jenny was used in the home, multiplying the work of one spinner manyfold, and for the first time making indoor work more lucrative than outdoor work for much of the population. By allowing both men and women to work within the home, it helped cement the nuclear family, provided a better working environment for children, and broke the dependency on landowners. It was also a way for regular people to become entrepreneurs without having to go through the apprentice process of the guild system. Even as factories grew around the cottages, that sort of domestic entrepreneurship remained popular as a way for companies to outsource piecework to a network of highly skilled artisans whose output was multiplied by micro-manufacturing techniques.

The spread of these machines marked the end of the mostly agrarian era of British history. Rather than most people working in the fields, fewer people with better farming machines could plow and harvest, while the rest worked in the home in domestic workshops, with spinning soon joined by weaving and knitting with wooden looms.

Because such work wasn't tied to the land, it wasn't tied to landowners, either. The family members working in the home had more independence and control over their own economic future. But though they were liberated from a single landowner, they now had to deal with the market forces of supply and demand. They sold to big industrial buyers who were always seeking lower prices and would shift their buying to get them.

Wages were often no better than in farmwork, but at least the workers could set their own schedule. It was a step toward entrepreneurialism, but it fell short of creating truly differentiated innovation. Instead, most cottage industries were simply distributed labor for the big factories, compensating for their inferior machines by not requiring the factories to make capital investments in new production equipment or retooling for small or unusual orders. It was thatched-roof manufacturing, but not thatched-roof invention. The cottage workers were always at the mercy of the industrialists.

Nevertheless, the rise of cottage industries was an important part of the First Industrial Revolution that is often overshadowed by the image of the big "dark satanic mills." In a sense, they were closer to what a Maker-driven New Industrial Revolution might be than are the big factories we normally associate with manufacturing. Cottage industries were a distributed form of production, which complemented the centralized factories by being more flexible and making things in smaller batches than the big factories could gear up for.

They fit into and reinforced the family structure, finding work for all the family members (including, like it or not, lots of children, contributing to the population explosion that defined that period of British history). While big factories were drawing young adults to the cities to work and live in industrial compounds, cottage industries were growing the market towns. And they emphasized and preserved prized artisanal skills such as lacework, which at the time were difficult for machines or otherwise commanded a premium price.

Cottage industries were a thriving market well into the nineteenth century. In the late 1830s, for example, Dixons of Carlisle employed 3,500 handloom weavers scattered around neighboring counties, and a decade later Wards of Belper was recorded as providing work for four thousand scattered knitting frames. As late as the 1870s, Eliza Tinsley and Co. was putting out work to two thousand cottage nail and chain makers in the British Midlands.[18] Even at the height of the First Industrial Revolution, the distributed labor of cottage industries ensured that there were far more small businesses than large ones.

Compare that with a typical Maker-ish small company today. Today's cottage industry is more typically an Etsy marketplace seller with a computer-controlled vinyl cutter making cool stickers for Macbooks or making and selling perfect replacement parts for vintage cars. Like their Industrial Age ancestors, they typically make the kinds of things big factories do not—they focus on niche markets of thousands, not mass markets of millions. They're distributed in a way that reflects the natural geography of ideas, not the hub-and-spoke logic of massive supply chains and cheap industrial land.

They're often run out of the Maker's garage or workshop, at least at the start, and often use family members as help. They make a virtue of their small-batch status, emphasizing handcrafted or artisanal qualities. And they are focused on desktop production tools, best suited for hundreds or a few thousand pieces.

That speaks to another key principle of the Maker Movement: As with the spinning jenny over two hundred years ago, the technology to create and design new products is available to anyone today. You don't need to invest in a massively expensive plant or acquire a vast workforce to turn your ideas into reality. Manufacturing new products is no longer the domain of the few, but the opportunity of the many.

Rather than selling to factories that control the path to market, today's Maker-style cottage industries sell directly to consumers around the world online, on their own websites or through marketplaces like Etsy or eBay. Rather than wait for orders from factories, as their nineteenth-century ancestors did, they invent their own products and seek to build their own microbrands. And rather than competing on price in a commodity market that favors cheap labor, they compete on innovation. They invent their own designs and can charge a premium to their discriminating consumers who are intentionally avoiding mass-produced goods.

So, back to the future. Today we are seeing a return to a new sort of cottage industry. Once again, new technology is giving individuals the power over the means of production, allowing for bottom-up entrepreneurship and distributed innovation. Just as the Web's democratization of the means of production in everything from software to music made it possible to create an empire in a dorm room or a hit album in a bedroom, so the new democratized tools of digital manufacturing will be tomorrow's spinning jennies. And the guilds they may break may be the very factory model that grew up in Manchester and dominated the past three centuries.

Chapter 4

We Are All Designers Now

So we might as well get good at it.

When I was in high school in the late 1970s, we had workshop class as part of the "Industrial Arts" curriculum. It wasn't quite clear why this was a required credit—we lived in a suburb of Washington, D.C., and there were no factories around and most of my friends' parents were lawyers and government workers. But learning how to use workshop tools—band saws, table saws, drill presses, and the like—was just part of a mid-twentieth-century American education. The bad kids made ninja throwing stars; the worst made bongs. I made a crude magazine stand that my parents tolerated until I left home; I was lucky to have kept all my fingers through the process. Meanwhile, girls were steered to "Home Economics" to learn about sewing, cooking, and planting, which was, in a sense, another form of required crafting and DIY education.

At home, I made Heathkit electronics kits, which involved soldering irons and weeks of painstaking work with wires and components but were the cheapest way to obtain something like a citizen's band radio or a stereo amplifier. Chemistry kits had actual chemicals in them (as opposed to little more than baking soda and a ream of legalistic warnings, as is now sadly the case), and were great fun. Anybody with a cool or temperamental car spent the weekend under the hood with a wrench, hopping it up and otherwise tinkering with its mechanics. "Taking things apart to see how they work" was just what kids did, and finding uses for the parts launched countless fantastic machines, some of which actually worked.

But starting in the 1980s and 1990s, the romance of making things with your own hands started to fade. First, manufacturing jobs were no longer a safe way to enter and stay in the middle class, and the workshop lost even its vocational appeal as the number of manufacturing workers in the employment rolls shrank. In its place came keyboards and screens. PCs were introduced, and all the good jobs used them; the school curriculum shifted to train kids to become "symbolic analysts," to use the social-science phrase for white-collar information workers. Computer class replaced shop class. School budget cuts in the 1990s were the nail in the coffin; once the generation of workshop teachers retired, they were rarely replaced; the tools were sold or put in storage.

Imported Asian electronics became better and cheaper than Heathkit gear, and the shift from individual electronic components like resistors and transistors and capacitors to inscrutable microchips and integrated circuits made soldering skills pointless. Electronics became disposable boxes with "no user serviceable parts inside," as the warning labels put it. Heathkit left the kit business in 1992.[19]

Cars evolved from carburetors and distributor caps that you could fiddle with to fuel injection and electronic ignition that you couldn't. Chips replaced mechanical parts. The new cars didn't need as much maintenance, and even if you wanted to go under the hood there wasn't much you could fix or modify, other than to change the oil and the oil filter. The working parts were hermetically sealed and locked down, a price we happily paid for reliability and minimal upkeep.

Just as shop class disappeared with school budget cuts, better opportunities in the workplace for women and gender equality killed Home Economics. Kids grew up with computers and video games, not wrenches and band saws. The best minds of a generation were seduced by software and the infinite worlds to be created online. And they made the digital age we all live in today.

That is how the world shifted from atoms to bits. The transformation has gone on for thirty years, a generation, and it's hard to argue with any of it.

But now, thirty years after "Industrial Arts" left the curriculum and large chunks of our manufacturing sectors have shifted overseas, there's finally a reason to get your hands dirty again. As desktop fabrication tools go mainstream, it's time to return "making things" to the high school curriculum, not as the shop class of old, but in the form of teaching *design*.

Today, schoolchildren learn how to use PowerPoint and Excel as part of their computer class, and they still learn to draw and sculpt in art class. But think how much better it would be if they could choose a third option: design class. Imagine a course where kids would learn to use free 3-D CAD tools such as Sketchup or Autodesk 123D. Some would design buildings and fantastic structures, much as they sketch in their notebooks already. Others would create elaborate videogame levels with landscapes and vehicles. And yet others would invent machines.

Even better, imagine if each design classroom had a few 3-D printers or a laser cutter. All those desktop design tools have a "Make" menu item. Kids could actually fabricate what they have drawn onscreen. Just consider what it would mean to them to hold something they dreamed up. This is how a generation of Makers will be created. This is how the next wave of manufacturing entrepreneurs will be born.

"Desktop" changes everything

Two decades after desktop publishing became a mainstream reality, the word *desktop* is being added to industrial machinery, with equally mind-blowing effect. Desktop 3-D printing. Desktop computer-controlled routing, milling, and machining. Desktop laser cutting. Desktop computer-controlled embroidering, weaving, and quilting. Even desktop 3-D scanning, or "reality capture," digitizing the physical world. Desktop fabrication is leading to full-on desktop manufacturing.

What's so important about the word *desktop*? Just consider the history of the computer itself. Until the late 1970s, *computing* connoted room-sized mainframes and refrigerator-sized minicomputers, which were the sole domain of governments, big companies, and universities. Technologists had long predicted that computing would find a place in the average home—the Moore's Law trend of declining price and increasing power practically guaranteed that day would eventually come. But they couldn't imagine why anyone would want one. Computing was then used for tabulating census results and company accounting, running scientific simulations, and designing nuclear weapons—big, serious number-crunching. What need did a home have for that?

Companies from IBM to AT&T's Bell Labs got their best minds to brainstorm how a computer would be used in the future home, and came up with precious little. The most common prediction was that it would be used for recipe-card management. Indeed, in 1969 Honeywell even offered a $10,000 "kitchen computer" (official name: the "H316 Pedestal Model"), which was promoted on the cover of the Neiman-Marcus catalog to do just that—it was stylishly designed, with a built-in cutting board. (There is no evidence that any actually sold, not least because the very modern cook would have to enter data with toggle switches and read the recipes displayed in binary blinking lights.)

Yet when the truly personal—"desktop"—computer did eventually arrive with the Apple II and then the IBM PC, countless uses quickly emerged, starting with the spreadsheet and word processor for business and quickly moving to entertainment with video games and communications. This was not because the wise minds of the big computer companies had finally figured out why people would want one, but because people found new uses all by themselves.

Then, in 1985, Apple released the LaserWriter, the first real desktop laser printer, which, along with the Mac, started the desktop publishing phenomenon. It was a jaw-dropping moment, combining in the public imagination words that had never gone together before: "desktop" and "publishing"! Famously, Apple's printer had more processing power than the Mac itself, which was necessary to interpret the Postscript page description language that was originally designed for commercial printers costing ten times as much. But Steve Jobs wanted the Mac desktop publishing suite not just to match the quality of commercial printers, but to exceed them. Desktop tools could be *better* than traditional industrial tools, he believed, and he started by cutting no corners. (As a result, the printer launched at a relatively high price of $7,000 and required the invention of a new network technology so that many people in a small office could share it.)

Remember, at that time publishing used to mean manufacturing in every sense of the word, from the railways that brought huge rolls of paper and barrels of ink to the printing plant to the fleets of trucks that took the finished goods to market. The "power of the press" came from the massive printing presses of that era; the newspaper unions that still exist today are a reminder that newspapers used to be factories with blue-collar workers pushing pallets of paper around.

But with desktop publishing, a smaller version of this was within reach of anyone. In a sense, you could "prototype" a publication at home by printing a few copies, and then, when it looked right, you could take a floppy disk with the file to a copy shop to be printed in volume. Consumer-grade desktop tools spoke the same language (Postscript) as the biggest industrial printing plants. It wasn't for everyone at the start, to be sure, but over time, high-quality color desktop printers got cheaper and better. Today such printers cost less than $100 and are in practically every home (the killer app turned out to be digital photography, not newsletters and flyers).

Taking publishing out of factories liberated it. But the real impact of this was not in paper, but in the idea of "publishing" online. Once people were given the power of the press, they wanted to do more than print out newsletters. So, when the Web arrived, "publishing" became "posting" and they could reach the world.

Even the simple act of posting online is a way of occupying what were once factories. Today your PC is seamlessly connected to warehouse-sized server farms (the "cloud") that allow you to access massive-scale computing in an instant. You may not think of a simple Google search as an act of harnessing industrial-class computing, but until a few decades ago you'd have needed access to a multimillion-dollar supercomputer to search that much data. And if you've ever seen one of Google's server farms, you'll know that the factory comparison is not far off—they are the size of a city block. Now these are open to all to publish or retrieve their every notion globally, for free.

So there you have it: the industrial machinery of the biggest twentieth-century media empires transformed into the sort of thing that you can command from your laptop. Yesterday the biggest computing facilities in the world were working for the government, huge companies, and research labs. Today they're working for you. That is what the "desktop" wrought.

DIY design

So now the 3-D printer is where Jobs's Macintosh and LaserWriter were twenty-five years ago. As with the first laser printers, 3-D printing is still a bit expensive and hard to use; it's not yet for everyone. We haven't really figured out what the killer app will be. But what we do know is that it will get better and cheaper even faster than the laser printer did, thanks to all the basic mechanical and electronic technology 3-D printing shares with its dimensionally deprived ancestor, the inkjet printer you've already got. The only real difference is that it squirts a different liquid (molten plastic, not ink) and has one more motor to control height.

Like then, the first users are a little lost. When desktop publishing was first introduced, tens of thousands of people discovered that they knew nothing about fonts, kerning, text flow, anchors, and all that; they had to learn a couple of centuries' worth of publishing terms and techniques overnight. Many garish documents with a dog's breakfast of typefaces ensued, but so did an explosion of creativity that ultimately led to today's Web.

Today, with the spread of desktop fabrication tools, a generation of amateurs is also being suddenly confronted with the baffling language and techniques of professional industrial design, just as they were in the desktop publishing era. Now it's not wraparound text and line justification, but "meshes" and "G-code," "rasters" and "feedrates." Don't worry—you'll know what you need to soon enough, and someday kids will be taught these skills in fifth-grade digital fabrication class. Remember, the early days of the personal computer revolution were equally arcane—"pixels," "bytes," "RAM"—and now we hardly give the details of computing a thought, in part because maturing technology hides most of that plumbing from us.

So, too, for the Maker Movement. Today it is full of people who are dazzled by the potential of the industrial-quality tools now appearing on their desktops. The alien language and techniques of physical creation are intoxicating for the geeks; they're rushing to explore this strange new world. But that is just the first wave of what is quickly becoming a mainstream phenomenon. Soon these early tools will become as ubiquitous and as easy to use as inkjet printers. And if history is any guide, it will change the world even faster than the microprocessor did a generation ago.

We are all designers now. It's time to get good at it.

Chapter 5

The Long Tail of Things

**Mass production works for the masses.
But what works for you?**

One recent Saturday, my two youngest daughters decided they wanted to redecorate their dollhouse. They've been playing The Sims 3, which is a video game that's basically a virtual dollhouse where you can make any kind of home with a dizzying array of furniture and people choices ("Sims"), and then watch them live their lives in it. One daughter did her Sims house in modern "career girl" style, with a home gym and AV room. The other went more 1960s style, with stream-lined appliances, mod furniture, and an angular swimming pool.

Once their "screen time" was over, they wanted to continue play-ing out the theme with their real dollhouse. This is a sign of children brought up in the digital world, where anything is possible and every-thing is available. There are hundreds of furniture options available in The Sims. Why settle for anything less in the physical world?

But things don't always work that way in real life. Or at least not yet.

Their first instinct, of course, was to come to me and ask me to buy new furniture for them. And my own first instinct (after saying "no" and "wait for your birthday") was to at least find out what was avail-able. I went online and quickly realized three things: (1) dollhouse furniture is expensive; (2) there is surprisingly little variety; and (3) the stuff your kids like is invariably the wrong size for your dollhouse. Sorry, girls.

At that point, to my delight, they asked if we could make the furniture ourselves. My pleasure in their DIY spirit was slightly tempered, however, by memories of how projects started together with kids typically end up hours later with Dad in the workshop alone cursing broken bits of wood and X-Acto knife cuts. And even if I were to persevere, a week-long process of micro-carpentry would probably end up, if history is any guide, with my clumsy bit of misshapen wood being placed in the dollhouse's attic, unable to compete with the store-bought stuff on the other floors.

But now we have a 3-D printer, a MakerBot Thing-O-Matic, and so this quest ended differently. We went to Thingiverse, an online repository of 3-D designs that people have uploaded. And there it was, just like The Sims. Every furniture type we could want, from French Renaissance to *Star Trek*, was available, ready for the downloading. We grabbed some exquisite Victorian chairs and couches, resized them with a click to perfectly fit our dollhouse scale, and clicked on "build." Twenty minutes later we had our furniture. It was free, fast, and there was so much more choice than in the real world, or even on Amazon. We may never buy dollhouse furniture again.

If you're a toy company, this story should give you chills.

As I was writing this, Kodak went into bankruptcy, a victim of the shift away from film that needed to be bought and processed to digital photography, which is free and can be printed at home on desktop inkjet printers. If you're making cheap plastic toys today, can you see a premonition of your future in that?

Of course, physical objects are more complex than 2-D images. Right now we can print plastic in only a few colors on our MakerBot. The finish is not as good as injection-molded plastic, and we can't print color details with nearly as fine precision as the painting machines or stencils of Chinese factories.

But that's because we're at the dot-matrix equivalent of 3-D printers. Remember them, from the 1980s? They were noisy, monochrome, and crude—tiny pins hitting a black ink ribbon, little more than an automated electric typewriter. But today, just a generation later, we have cheap and silent inkjets that print in full color with resolution almost indistinguishable from professional printing.

Now fast-forward the clock a decade or two from today's early 3-D printers. They will be fast, silent, and able to print a wide range of materials, from plastics to wood pulp and even food. They will have multiple color cartridges, just like your inkjet, and be able to print in as many color combinations. They will be able to print images on the surface of an object even finer than the best toy factories today. They may even be able to print electronic circuits right into the object itself. Just add batteries.

Disruptive by design

Transformative change happens when industries democratize, when they're ripped from the sole domain of companies, governments, and other institutions and handed over to regular folks.

We've seen this picture before: it's what happens just before monolithic industries fragment in the face of countless small entrants, from the music industry to newspapers. Lower the barriers to entry and the crowd pours in.

That's the power of democratization: it puts tools in the hands of those who know best how to use them. We all have our own needs, our own expertise, our own ideas. If we are all empowered to use tools to meet those needs, or modify them with our own ideas, we will collectively find the full range of what a tool can do.

The Internet democratized publishing, broadcasting, and communications, and the consequence was a massive increase in the range of both participation and participants in everything digital—the Long Tail of bits.

Now the same is happening to manufacturing—the Long Tail of things.

My first book, *The Long Tail*, was about exactly this—the shift in culture toward niche goods—but mostly in the digital world.

For most of the past century, the natural variation and choice in products such as music, movies, and books have been hidden by the limited "carrying capacity" of the traditional distribution systems of physical stores, broadcast channels, and megaplex movie theaters. But once these products were available online in digital marketplaces with unlimited "shelf space," for lack of a better phrase, demand followed: the monopoly of the blockbuster was over. The mass market in culture has turned into a Long Tail of micro-markets, as any contact with a teenager these days will confirm (we're all indie now!).

In short: our species turns out to be a lot more diverse than our twentieth-century markets reflected. The limited store selection of our youth reflected the economic demands of retail of the day, not the true range of human taste. We are all different, with different wants and needs, and the Internet now has a place for all of them in the way that physical markets did not.

That was not exclusively digital, of course. The Internet also lengthened the tails of physical product markets for consumers. But it did so by revolutionizing *distribution*, not production.

For physical goods, the twentieth-century limits to choice were based on three distribution bottlenecks—you could only buy things that passed all of the three tests:

1. The products were popular enough for manufacturers to make.

2. The products were popular enough for retailers to carry.

3. The products were popular enough for you to find (via advertising or prominent placement in stores near you).

As Amazon showed, the Web could help with the latter two, right out of the gate.

First, Amazon and others who used centralized distribution warehouses and, later, distributed warehousing in the form of listings for third-party merchants who handle all the fulfillment, meant that they could list many more products than any physical retailer could carry. (Like the original catalog retailers, but without the limited pages of a paper catalog sent through the mail.)

Second, the shift to search as a discovery mechanism meant that people could find products that were not necessarily popular enough to promote the usual way in bricks-and-mortar retail.

Meanwhile, eBay did the same thing for used goods, countless specialist Web retailers emerged, and eventually Google aggregated them all into the ultimate way to find anything. Today the Web has already surfaced a Long Tail of products to rival the tail of digital ones. Bottlenecks 2 and 3 above are largely removed.

What about the first bottleneck—making more variety in the first place? Well, the Web helped some there, too. Its ability to tap "diffuse demand" (which is to say, products that aren't popular enough in any one place to carry in physical stores, but make sense when you can aggregate demand from around the world) meant that manufacturers could find markets for goods that otherwise would fail the test of traditional distribution. So more niche products were made, because they could find sufficient demand in selling to a global market online.

But that was just the start. Remember that the real Web revolution was not that we could just buy more stuff with greater choice, but *make our own stuff* that others could consume. The spread of digital cameras meant an explosion of videos that YouTube could distribute, and digital desktop tools did the same for music, publishing, and software creation. Anybody could make anything, given enough talent. Access to powerful tools and a means of distribution was no longer a barrier to participation. If you had talent and drive, you could find an audience, even if you didn't work for the right company or have the right degree.

In the Web case, the "stuff" was and is mostly creativity and expression in digital form: words, pictures, videos, and the like. It doesn't compete with commercial goods for money, but does compete for time. A blog may not be a book, but at the end of the day, it's just another way to entertain and inform. The greatest change of the past decade has been the shift in time people spend consuming amateur content instead of professional content. The rise of Facebook, Tumblr, Pinterest, and all the others like them is nothing less than a massive attention shift from the commercial content companies of the twentieth century to the amateur content companies of the twenty-first.

Now the same is happening with physical goods. The 3-D printers and other desktop prototyping tools are the equivalent of the cameras and music editing tools. They allow anyone to create one-offs for their own use. As Rufus Griscom, a Web entrepreneur who founded Babble.com, puts it, "this is the Renaissance of Diletantism."

At the same time, the world's factories are opening up, offering Web-based manufacturing as an on-demand service to anyone with a digital design and a credit card. They allow a whole new class of creators to go into production, turning their prototype into a product, without having to build their own factories or even have companies themselves. Manufacturing has now become just another "cloud service" that you can access from Web browsers, using a tiny amount of vast industrial infrastructure as and when you need it. Somebody else runs these factories; we just access them when we need them, much as we can access the huge server farms of Google or Apple to store our photos or process our e-mail.

The academic way to put this is that global supply chains have become "scale-free," able to serve the small as well as the large, the garage inventor *and* Samsung. The non-academic way to say it is this: nothing is stopping you from making anything. The people now control the means of production. Or, as *The Lean Startup* author Eric Reis puts it, Marx got it wrong: "It's not about ownership of the means of production, anymore. It's about *rentership* of the means of production."

Such open supply chains are the mirror of Web publishing and e-commerce a decade ago. The Web, from Amazon to eBay, revealed a Long Tail of demand for niche physical goods; now the democratized tools of production are enabling a Long Tail of supply, too.

The industrial artisan

The Long Tail of things is around you already and has been for years, just not on this scale. Take any domain where you have a deep interest and start searching online. Got a classic car, perhaps an old MG roadster? A few clicks in your browser and you're in the domain of hyperspecialized suppliers who focus on making nothing but replacement bonnet release cables for car models that haven't been made for a generation. Or perhaps you're looking for a jewelry tree on which to dangle necklaces. You may start at Crate & Barrel, but five clicks later and you're on Etsy, buying something much cooler and more interesting (and no more expensive) from a metal artist in Texas. The barriers to variety have disappeared.

Now the rise of the "artisanal" movement and mass-scale crafting has created widespread demand for such specialized goods. There is, as I write, a glut of artisanal pickle makers in Brooklyn. Meanwhile, the artisanal mustard market is booming here in Berkeley; even Wal-Mart now sells more than a hundred kinds of mustard, including scores of natural stoneground varieties. The local chocolate makers, such as Tcho, compete on which has the deepest, most ethical supply chain. It's one thing to say you're "organic" and "fair trade," but do you start with the actual beans? And buy them straight from Ghana? And know the names of some of your pickers? For people who care about such things, it's hard to beat the artisans for sheer obsessive *caring* about what they do.

What's different about these niche physical goods, created by people and communities who aren't attempting to conform to the economic requirements of Big Manufacturing?

For starters, niche goods aimed at discriminating audiences can command higher prices. Just think of couture fashion or fine wines. Boutique products with unique qualities are polarizing—they may be just right for you but not for others. But the people they really *are* for are often willing to pay more for the privilege of being so well suited. From tailored clothes to fancy restaurants, exclusivity has always commanded a premium.

This is what i.materialize, a design firm, calls "the power of the unique." In a world dominated by one-size-fits-all commodity goods, the way to stand out is to create products that serve individual needs, not general ones. Custom-made bikes fit better. Right now this is mostly the privilege of the rich, as such products require handcrafting. But what if they could be produced using digital manufacturing where there is no cost to complexity and no penalty for short production runs?

Increasingly, when computers are running the production machines, it costs no more to make each product different. If you've ever received a catalog or magazine in the mail that has a personalized message for you, that's a formerly one-size-fits-all production machine—the printing press—turned into a digital one-size-fits-one machine, using little more than a big version of the desktop inkjet printer. Likewise when you buy a cake with fancy icing from the supermarket. That icing was applied by a robot arm—it can make each cake design different as quickly as making them all the same—personalizing a cake costs no more to do, yet the supermarket can charge more for it because it is perceived as more valuable. The old model of expensive custom machines that had to make the same thing in vast numbers to justify the tooling expense is fading fast.

These niche products tend to be driven by *people's* wants and needs rather than companies' wants and needs. Of course people have to create companies to make these goods at scale, but they work hard to retain their roots. Such entrepreneurs often state that their first obligation is to serve their community, and to make money second. Goods made by passionate consumers-turned-entrepreneurs tend to radiate a quality that displays craftsmanship rather than mass-manufactured efficiency.

In a sense, this is just the extreme of the specialization that Adam Smith originally recognized in *The Wealth of Nations* as the key to an efficient market. People should do only what they do best, he said, and trade with others who make other specialized goods. No one person or town should try to do it all, since a society can do far more collectively with an efficient division of labor—comparative advantage plus trade equals growth. What was good in the eighteenth century is even better in the twenty-first, now that specialists have access to global supply chains for their commodity input materials and global consumer markets for their niche output products.

Nearly thirty years ago, two MIT professors, Michael Piore and Charles Sabel, predicted this transition in a book titled *The Second Industrial Divide*. They argued that the mass-production model that defined twentieth-century manufacturing economies (the "first industrial divide" between people and production) was neither inevitable nor the end of innovation in making things.

> Under somewhat different historical conditions, firms using a combination of craft skill and flexible equipment might have played a central role in modern economic life—instead of giving way, in almost all sectors of manufacturing, to corporations based on mass production. Had this line of mechanized craft production prevailed, we might today think of manufacturing firms as linked to particular communities rather than as the independent organizations that, through mass production, seem omnipresent.[20]

Today, digital desktop fabrication has indeed introduced a sort of "mechanized craft production" that Piore and Sabel could only have dreamed of. Rather than returning to the sewing machines and local machine shops that big factories drove out of the market a hundred years ago, the modern Maker Movement is built on high-tech digital fabrication, and can let regular people harness big factories at will to make what they want. It's the perfect combination of inventing locally and producing globally, serving niche markets defined by taste, not by geography. And what's clear about these new producers is that they're not going to be making the same one-size-fits-all products that defined the mass-production era. Instead, they're going to be starting with one-size-fits-one and building from there, finding out how many other consumers share their interests, passions, and unique needs.

Happiness economics

What's interesting is that such hyperspecialization is not necessarily a profit-maximizing strategy. Instead, it is better seen as *meaning*-maximizing. Writing in *The New York Times Magazine*, Adam Davidson sees this as a natural evolution of an affluent country where the basic needs for the middle class and above have all been more than met:

> The hot field of happiness economics argues, rather persuasively, that once people reach some level of comfort, they are willing— even eager—to trade in potential earnings at a lucrative but uninspiring job for less (but comfortable) pay at more satisfying work. Some research by the Chicago economist Erik Hurst suggests that half of entrepreneurs start businesses as much to pursue happiness as to make money.[21]

What's more, consumers tend to value more highly products in which they feel they have had a hand in their creation, whether assembling a kit or just encouraging the creators themselves online. Researchers call this "the IKEA Effect," and it dates all the way back to the Home Economics movement. As Duke University behavioral economist Dan Ariely and his colleagues write in a paper on this,

> When instant cake mixes were introduced in the 1950s as part of a broader trend to simplify the life of the American housewife by minimizing manual labor, housewives were initially resistant: the mixes made cooking too easy, making their labor and skill seem undervalued. As a result, manufacturers changed the recipe to require adding an egg; while there are likely several reasons

why this change led to greater subsequent adoption, infusing the task with labor appeared to be a crucial ingredient.[22]

Today, in experiments with IKEA furniture, when the paper's study participants were given the opportunity to buy IKEA furniture they built themselves versus identical units built by others, they bid 67 percent more for their own creations. They did the same with Lego kits and paper origami. In all cases, people would pay more for things where their own sweat was one of the ingredients. This is the Maker's Premium. It's the ultimate antidote to commodification.

Take any niche and check out the new producers. Mountain bike parts, classic car accessories, cool vinyl "skins" for phones and other gadgets—they're all seeing a wave of new micro-entrepreneurs selling online. Although each market is different, what's common about this new creative class is that they were once consumers who wanted something that didn't exist before. So, rather than settle for what was on the market, they made something better themselves. And once they made one, it was increasingly easy to make more. And thus a small business emerged from the most passionate ranks of the consumer class.

What does *artisanal* mean in a digital world? In his 2011 book, *The Alphabet and the Algorithm*, Mario Carpo, an Italian architectural historian, argues that "variability is the mark of all things handmade." So far, no surprise for anyone who has bought a tailored suit. But he continues:

> Now, to a greater extent than was conceivable at the time of manual technologies . . . the very same process of differentiation can be scripted, programmed, and to some extent designed. Variability can now become part of an automated design and production chain.[23]

Just consider the Web itself. Each of us sees a different Web. When we visit big Web retailers such as Amazon, the storefront is reorganized just for us, displaying what its algorithms think we'll most like. Even for pages where the content is the same, the ads are different, inserted by software that evaluates our past behavior and predicts our future actions. We don't browse the Web, but rather search it, and not only are our search strings different, but different users get different results from the same search strings based on their personal history.

Writes Carpo, "This is, at the basis, the golden formula that has made Google a very rich company. Variability, which could be an obstacle in a traditional mechanical environment . . . has been turned into an asset in the new digital environment—indeed, into one of its most profitable assets."

Information inside

But surely custom-made or bespoke suits and farmers' markets have been around forever. What's different now? The simple answer is that DIY culture has suddenly met Web culture. And the intersection of the two lies in digital design: physical products that are created first onscreen.

Walk into an Apple Store and look around you. All those shiny objects—all those beautifully designed and manufactured slabs of titanium, high-end plastics, and circuitry—started life on a screen somewhere. So, too, for a Nike store. Or a car dealership.

Physical products are increasingly just digital information put in physical form by robotic devices such as CNC mills and pick-and-place machines making printed circuit boards. That information is a design, translated into instructions to automated production equipment. In a sense, hardware is mostly software these days, with products becoming little more than intellectual property embodied in commodity materials, whether it's the code that drives the off-the-shelf chips in gadgets or the 3-D design files that drive manufacturing.

And the more products become information, the more they can be treated as information: collaboratively created by anyone, shared globally online, remixed and reimagined, given away for free, or, if you choose, held secret. In short, the reason atoms are the new bits is that they can increasingly be made to *act like bits*.

The result is that we are now seeing what looks like what Joseph Flaherty, who writes the Replicator blog, calls a "Moore's Law for Atoms." The original Moore's Law, named after Intel researcher Gordon Moore, described the twenty-four-month doubling of processing power per dollar that has characterized the computer industry since the 1970s. That exponential growth comes from the phenomenon of "compound learning curves": breakthrough discoveries in semiconductor research come frequently enough (about every three years) and build on their predecessors so effectively that progress accelerates at this breakneck pace.

Why do all industries not enjoy this pace of improvement? Because semiconductors are still a relatively new field in the long arc of scientific research. They are built on the quantum mechanics and material science breakthroughs of the early twentieth century, a remarkable period of discovery that opened an entirely new domain of physics. As Richard Feynman famously said, "there's a lot of room at the bottom," at the atomic level of matter, and we're still just beginning to plumb it.

What is the analogy for manufacturing? Nothing so grand as a new physics. Instead, it is simply the combination of the technologies that the original Moore's Law brought us: computers, digital information, the Internet, and, most important of all, connected people.

Remixing the physical world

It's easy to miss the magnitude of this shift. After all, from a distance the whole process of making things doesn't seem so different. My grandfather designed his machines on paper and prototyped them by hand in his workshop. I design in CAD and send the files to be prototyped on my desktop fabricator or by robot machines in a remote service bureau. But at the end of the process, we still both have a prototype in hand. What's the big deal in doing it my way?

The answer comes down to the unique qualities of digital information. It seems like such a small distinction: products shared as physical things or products shared as *digital descriptions* of physical things. After all, if you've got to fabricate them to make them real, one way or another, who cares what form the instructions take?

But as we've learned over the past few decades, digital is different. Sure, digital files can be shared and copied limitlessly at virtually no cost and with no loss of quality. But what's more important is that they can be *modified* just as easily. We live in a "remix" culture: everything is inspired by something that came before, and creativity is shown as much in the reinterpretation of existing works as in original ones. That's always been true (the Greeks argued that there were only seven basic plots, and all stories just changed the details of one or another of them), but it's never been easier than it is now. Just as Apple encouraged music fans to "Rip. Mix. Burn," Autodesk now preaches the gospel of "Rip. Mod. Fab" (3-D scan objects, modify them in a CAD program, and print them on a 3-D printer).

That ability to easily "remix" digital files is the engine that drives community. What it offers is an invitation to participate. You don't need to invent something from scratch or have an original idea. Instead, you can participate in a collaborative improvement of existing ideas or designs. The barrier to entry of participation is lower because it's so easy to modify digital files rather than create them entirely yourself.

My grandfather was a lone inventor, not because he was especially solitary but because he had no mechanism for easy sharing. I may be no more extroverted than he was, but because my medium is digital, sharing comes naturally. When you share, community forms. And what community does best is remixing—exploring variation in what a product can be, and in the process improving it and propagating it far faster than any individual or single company could.

Think of a digital product design not as a picture of what it should be, but instead as a mathematical equation of how to make it. That is not a metaphor—it's actually the way CAD programs work. When you draw a 3-D object on the screen, what the computer really does is write a series of geometrical equations that can instruct machines to reproduce the object at any size in any medium, be it pixels on a monitor or plastic in a printer. Increasingly, those equations don't just describe the shape of a thing, but also its physical properties—what's flexible and what's stiff, what conducts electricity and what insulates heat, what's smooth and what's rough.

So everything is an algorithm now. And just as every Google search uses its algorithms to produce a different result for each person searching, so can algorithms customize products for their consumers.

For the *99 Teapots* project, Architect Greg Lynn designed one teapot in a CAD package, then let the software remix it to create ninety-eight others. Each was fabricated in a carbon mold, and titanium was exploded within it to create a unique teapot. (With a price tag of up to $50,000, they were more art piece than serving set, but the process was as interesting as the product.)

Lynn explained the point: this sort of variation in form is the essence of being a modern designer. In a 2005 speech at the TED Conference, he explained the BMW design challenge. At any given time, the car company has scores of designs, ranging from the $30,000 300 Series to the $70,000 700 Series. All of the BMW cars should "look like BMWs," which is to say there should be a family resemblance. But if the 700 Series is going to justify costing more than twice as much as a 300 Series, it can't look *too* much like it. Instead, it has to look more like other 700 Series cars.

What factors determine "BMW-ness"? And what factors determine "700 Series-ness"? It can't just be the mechanical specifications—there must also be some ineffable aesthetic that is hard to describe in words but easy to see. Decades ago, the ability to do that defined a master designer, and perhaps if you work for BMW and Apple, companies defined by personal design vision, it still does. But for most companies today, it defines a master algorithm instead. Software is increasingly driving the design process, with the broad strokes created by the human eye but the details and variations all filled in by code following rules dictated by material properties and manufacturing efficiencies, easily remixed by others into any number of variations.

Carpo explains what this represents: "Algorithms, software, hardware and digital manufacturing tools are the new standards of product design. . . . Unlike a mechanical imprint, which physically stamps the same form onto objects, an algorithmic imprint lets outward and visible forms change and morph from one object to the next."

Sound familiar? This echoes the "mass customization" promise of the first wave of Web retail, a decade ago. If a product is built on demand, why not have it designed on demand, too, or at least offer the consumer the ability to customize it according to taste? Dell's success with bespoke computers a decade ago promised an era when everything from cars to clothes would be made and sold that way.

But it didn't happen, at least not at the scale that everyone expected. Cars, for example, are chosen primarily for their reliability. The more variability there is in the manufacturing process, the harder it is to keep the defect rate down. Without perfect 3-D models of the customer (and telepathic understanding of customers' wearing preferences), clothes are hard to tailor predictably, which is why men still have their inseams measured in shops.

Today the canonical examples of mass customization are still a bit trivial, to say nothing of getting long in the tooth: Nike ID shoes (you can design a novel pattern on standard sneakers), custom-printed M&Ms, and the like. Having your name inscribed on the back of your iPad is hardly an industrial revolution.

And even Dell hardly does mass customization anymore. Today you can only choose the standard models with their two or three choices of memory, CPU, hard drive, and video card options, and if you don't pick the most popular combination (which Dell mass-produces the good old-fashioned way), you'll have to wait an extra two weeks for delivery. Car companies do the same. They all found that more variety meant more variability in quality and uncertainty in inventory. Given a choice between infinite options and products that are cheap, available, and reliable, consumers tended to go the safe, one-size-fits-all route.

Likewise, the examples where consumers are designing their own products online are rarely mass. Threadless (T-shirts), Lulu (self-published books), CafePress (coffee mugs and other trinkets), and others like them are thriving businesses, but they are platforms for creativity more than great examples of mass customization. They simply give consumers access to small-batch manufacturing on standard platforms: shirts, mugs, and bound paper.

So I won't be invoking "mass customization" much here. Instead, what the new manufacturing model enables is a *mass market for niche products*. Think ten thousand units, not ten million (mass) or one (mass customization). Products no longer have to sell in big numbers to reach global markets and find their audience. That's because they don't do it from the shelves of Wal-Mart. Instead, they use e-commerce, driven by an increasingly discriminating consumer who follows social media and word of mouth to buy specialty products online.

In a 2011 speech at Maker Faire, Neil Gershenfeld, the MIT professor whose book *Fab: The Coming Revolution on Your Desktop* anticipated much of the Maker Movement nearly a decade ago, described his epiphany like this:

> I realized that the killer app for digital fabrication is personal fabrication. Not to make what you can buy in Wal-Mart, but to make what you *can't* buy at Wal-Mart.
>
> This is just like the shift from mainframes to personal computers. They weren't used for the same thing—personal computers are not there for inventory and payroll. Instead personal computers were used for *personal* things, from e-mail to video games. The same will be true for personal fabrication.[24]

Small batches

Blogger Jason Kottke wrestled with what to call this new class of entrepreneurship, these cottage industries with global reach targeting niche markets of distributed demand. "Boutique" is too pretentious, and "indie" not quite right. He observed that others had suggested "craftsman, artisan, bespoke, cloudless, studio, atelier, long tail, agile, bonsai company, mom and pop, small scale, specialty, anatomic, big heart, GTD business, dojo, haus, temple, coterie, and disco business." But none seemed to capture the movement.

So he proposed "small batch," a term most often applied to bourbon. In the spirits world, this implies handcrafted care. But it can broadly refer to businesses focused more on the quality of their products than on the size of the market. They'd rather do something they were passionate about than go mass. And these days, when anyone can get access to manufacturing and distribution, that is actually a viable choice. Wal-Mart, and all the compromise that comes with it, is no longer the only path to success.

The collective potential of a million garage tinkerers is about to be unleashed on the global markets as ideas go straight into production, no financing or tooling required. "Three guys with laptops" used to describe a Web startup. Now it describes a hardware company, too. "Hardware is becoming much more like software," as MIT professor Eric von Hippel puts it.

The Web was just the proof-of-concept of what an open, bottom-up, collaborative industrial model could look like. Now the revolution hits the real world.

Part Two

The Future

Chapter 6

The Tools of Transformation

3-D printers are heading for the alchemist's dream: making anything.

"Tea. Earl Grey. Hot."

When Captain Jean-Luc Picard wants a steaming beverage in his ready room aboard the starship *Enterprise*, he just utters those words. The ship's "replicator" then assembles the necessary atoms—including those for the cup—and produces it, ready for the drinking. Picard thinks nothing of it—it's hardly more remarkable to him than a microwave oven is to us today. Just as we now use radio waves to excite atoms and generate heat in our own kitchens (which would have been mind-blowing in the 1950s), his replicator uses some fancy energy technology that is never quite specified in *Star Trek: The Next Generation* to get atoms to self-assemble into food and drink.

That's science fiction, but it's actually not impossible. When you see an industrial 3-D printer working today, with a little poetic license you can glimpse the beginnings of something similar. A bath of liquid resin lies inert, a primordial soup. A laser begins tracing pattens in it, like lightning. Shapes form and emerge from the nutrient bath, conjured as if by magic from nothing.

Okay, poetic license revoked—we're still a long way from molecular self-assembly, or at least in any useful way. A 3-D printer can work with only one material at a time, and if you want to combine materials you need to have multiple print heads or switch from one to another, like the different color cartridges in your desktop inkjet printer. We can only work at a resolution of about 50 micrometers (the thickness of a fine hair), while nature works at a thousand times finer detail, of a few tens of nanometers. And there's nothing self-assembling about the way a 3-D printer works: it does all the assembling itself, with the brute force of a laser solidifying a powder or liquid resin, or melting plastic and spreading it down in a fine line.

But you get the point. We can imagine something, draw it on a computer, and a machine can make it real. We can push a button and an object will appear (eventually). As Arthur C. Clarke put it, "any sufficiently advanced technology is indistinguishable from magic." This is getting close.

Four Desktop Factories

MakerBot Thing-O-Matic

1) 3-D PRINTER: A 3-D printer and the paper printer you've probably already got on your desktop play similar roles. The traditional laser (or inkjet) printer is a 2-D printer: it takes pixels on a screen and turns them into dots of ink or toner on a 2-D medium, usually paper. A 3-D printer, however, takes "geometries" onscreen (3-D objects that are created with the same sorts of tools that Hollywood uses to make CG movies) and turns them into objects that you can pick up and use.

Some 3-D printers extrude molten plastic in layers to make these objects, while others use a laser to harden layers of liquid or powder resin so the product emerges from a bath of the raw material. Yet others can make objects out of any material from glass, steel, and bronze to gold, titanium, or even cake frosting. You can print a flute or you can print a meal. You can even print human organs out of living cells, by squirting a fluid with suspended stem cells onto a support matrix, much as your inkjet printer squirts ink onto paper.

2) CNC MACHINE: While a 3-D printer uses an "additive" technology to make things (it builds them up layer by layer), a CNC (computer numerical control) router or mill can take the same file and make similar products with a "subtractive" technology, which is a fancy way of saying that it uses a drill bit to cut a

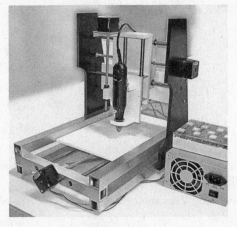

MyDIYCNC

product out of a block of plastic, wood, or metal. There are countless other specialty CNC machines: CNC quilters and embroidery machines, CNC sign and vinyl cutters (for silk-screening), and CNC paper and fabric cutters for crafters, to name a few. Some CNC machines are the size of a large table and are designed to make furniture out of wood (industrial CNC machines can be as big as a warehouse and can carve out objects as big as an airplane fuselage).

Epilog Zing laser cutter

3) LASER CUTTER: One of the most popular of the new desktop tools is the laser cutter, which is mostly a 2-D device. It uses a powerful laser to cut a precise pattern of any complexity into sheets of whatever material you feed it, from plastics and woods to thin metal. Many CAD programs can break a 3-D object into 2-D parts so they can be fabricated with a laser cutter, and then neatly slotted together like one of those plywood dinosaur kits.

Zscanner 3-D scanner

4) **3-D SCANNER:** This device, which can be as small as a bread-box, allows you to do "reality capture." Rather than having to draw an object from scratch, you can put an existing object in the scanner. It then uses lasers or other light sources and a camera to image the object from all sides, and then turns it into a 3-D image made up of tens or hundreds of thousands of polygons, just like a video-game character or CG movie set. The software can simplify it and let you modify any part you want. A common first experiment is to scan your head, then exaggerate your features and 3-D print a bobble-head of yourself.

You may think of 3-D printing as bleeding-edge technology today, the stuff of high-end design workshops and geeks. But you may have encountered a 3-D printer already, in ways so prosaic you didn't even notice.

Take custom dental fittings, such as those that change the alignment of the teeth over months with a series of slightly different mouth guards, each of which shifts the teeth imperceptibly into a new position. In that case, a dental technician scans the current position of your teeth; then software mathematically models all the intermediate positions to the desired endpoint. Finally, those positions are 3-D printed in plastic as a series of mouth guards that you wear, each for two or three weeks, until your teeth are in the new position.

Likewise for the prototypes of practically every gadget you've ever bought, and the architectural models for the newer buildings around you. Custom prosthetics are 3-D printed. If you're lucky enough to have a dentist who can replace a crown in a single sitting, that's probably 3-D printed (then sprayed with enamel) in the office. Doctors have printed and replaced an entire human jaw from titanium.

Today, you can buy a custom 3-D printed action figure of your World of Warcraft character or your Xbox Live avatar. And if you go to Tokyo, you can have your head scanned and you can buy a photorealistic action figure of yourself (try not to get too creeped out).

Commercial 3-D printing works with only a few dozen types of materials, mostly metals and plastics of various sorts, but more are in the works. Researchers are experimenting with more exotic materials, from wood pulp to carbon nanotubes, that give a sense of the scope of this technology. Some 3-D printers can print electrical circuits, making complex electronics from scratch. Yet others print icing onto cupcakes and extrude other liquid foods, including melted chocolate.

At the huge scale, there are already 3-D printers that can make a multistory building by "printing" concrete. Right now that requires a 3-D printer the size of the building, but it may someday be built into the cement truck itself, with a concrete that uses positional awareness to decide where to put down concrete and how much, directly reading and following the architect's CAD plans.

Meanwhile, researchers are working just as hard at moving in the other direction: 3-D printing at the molecular scale. Today there are "bio printers" that print a layer of a patient's own cells onto a 3-D printed "scaffold" of inert material. Once the cells are in place, they can grow into an organ, with bladders and kidneys already demonstrated in the lab. Print with stem cells, and the tissue will form its own blood vessels and internal structure.

Today's vision for 3-D printing is grand in ambition. Carl Bass, the CEO of Autodesk, one of the leading companies making 3-D authoring CAD software, sees the rise of computer-controlled fabrication as a transformative change on the order of the original mass production. Not only can it change the way traditional consumer goods are made, but 3-D printing can also work on scales as small as biology and as large as houses and bridges.

In an essay he published in the *Washington Post*, Bass explained what's so different about this way of making things:

> The ability to produce a small number of high-quality items and sell them at reasonable prices is causing an enormous economic disruption. In it, you can see the future of American manufacturing.
>
> In a computerized manufacturing process like 3-D printing, complexity and quality come at no cost. . . . A traditional paper printer can print a circle or a copy of the *Mona Lisa* with equal ease. The same rule applies to a 3-D printer.[25]

From a design perspective, this is revolutionary. It is no longer necessary for the designer to care or know about the manufacturing process, because the computer-controlled machines figure that stuff out for themselves. The same design can be fabricated in metal, plastic, cardboard, or cake icing. (It might not be very useful in all those materials, but it would exist.) "We can separate the design of a product from its manufacture for the first time in history, because all of the information necessary to print that object is built into the design," Bass explained.

Even better, as 3-D printers proliferate and are used for small-scale bespoke or custom-made manufacturing, they can provide a more sustainable way of making things. There are little or no transportation costs, because the product is made locally. There is little or no waste, because you use no more raw material than you need. And because the product is custom-made just for you, you're more likely to value it and keep it longer. Personalized products are less disposable; you simply care about them more.

Rich Karlgaard, the publisher of *Forbes* magazine, thinks that 3-D printing "could be the transformative technology of the 2015–2025 period." He writes:

This has the potential to remake the economics of manufacturing from a large-scale industry back to an artisan model of small design shops with access to 3-D printers. In other words, making stuff, real stuff, could move from being a capital intensive industry into something that looks more like art and software. This should favor the American skill set of creativity.[26]

But also remember what 3-D printing and any other digital production techniques cannot do. They offer no economies of scale. It is no cheaper on a per-unit basis to make a thousand than it is to make one. Instead, they offer exactly the opposite advantage: there is no penalty for changing each individual unit or making just a few of a kind.

It is the reverse of mass production, which favors repetition and standardization. Instead, 3-D printing favors individualization and customization. The big win of the digital manufacturing age is that we can have our choice between the two without having to fall back on expensive handcrafting: both mass and custom are now viable automated manufacturing methods.

If you want to make a million rubber duckies, you can't beat injection-molding. The first ducky may cost $10,000 in tooling for a mold, but every one after that amortizes the one-off cost. By the time you've made a million, they cost just pennies for the raw material. Make the same thing on a 3-D printer, on the other hand, and the first ducky might cost just $20 in time and materials—a huge savings. But so, sadly, will the one-millionth—there is no volume discount.

Include the amortized cost of the machine it takes to 3-D print those duckies one at a time (a process that might take an hour), rather than injecting-molding them in batches of a dozen or more at less than a minute per batch, and the crossover point where it's cheaper to go the injection-molding route comes at just a few hundred. For small batches, digital fabrication now wins. For big batches, the old analog way is still best (see diagram).

But just think about how many products actually make more sense

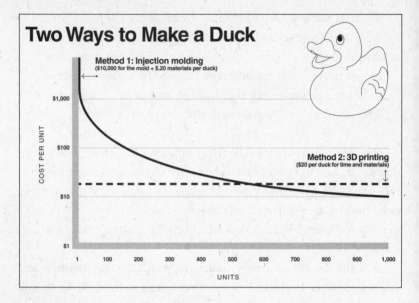

Two Ways to Make a Duck

Method 1: Injection molding
($10,000 for the mold + $.20 materials per duck)

Method 2: 3D printing
($20 per duck for time and materials)

COST PER UNIT

$1,000

$100

$10

$1

UNITS

1   100   200   300   400   500   600   700   800   900   1,000

in units of hundreds, not millions. For this Long Tail of things, the only option a few decades ago was handcrafting. But today digital fabricators can bring automated processes and near-perfect quality to the smallest batches. All those niche products that either weren't on the market at all because they didn't pass the economic test of mass production or were ruinously expensive because they needed to be handmade are now within reach.

Digital fabrication inverts the economics of traditional manufacturing. In mass production, most of the costs are in up-front tooling, and the more complicated the product is and the more changes you make, the more it costs. But with digital fabrication, it's the reverse: the things that are expensive in traditional manufacturing become free:

1. **Variety is free:** It costs no more to make every product different than to make them all the same.

2. **Complexity is free:** A minutely detailed product, with many fiddly little components, can be 3-D printed as cheaply as a

plain block of plastic. The computer doesn't care how many calculations it has to do.

3. **Flexibility is free:** Changing a product after production has started just means changing the instruction code. The machines stay the same.

You don't have to go to 3-D printing to see this in action. We already have this with a small class of familiar "standardized platforms" for customization: T-shirts and other simple clothing, coffee mugs, stickers, and the like. Companies such as Threadless, CafePress, and others have created huge businesses out of offering custom printing on such products. In this case, the enabling technology is not 3-D printing, but rather just 2-D printing on complex shapes and materials; the effect, however, is the same: a thriving market in the sort of products that would never make sense in a mass-production market.

Typically, Threadless and CafePress orders are in dozens—not ones, but not thousands, either. Yet collectively, this Long Tail can add up to a lot. CafePress has more than two million customers. In 2011, its revenues were $175 million. It is publically traded, and at the time of this writing was worth a quarter of a billion dollars.[27] Not bad for custom-printed T-shirts and mugs.

As easy as XYZ

Let's return to the 3-D printer, this miraculous machine that has so fired the imagination of futurists and machine-shop operators alike. How does it work?

At its core, a 3-D printer is just a variety of three-axis CNC machines. Two computer-controlled motors move a head left and right and forward and back (the x and y axis), while another motor moves the printer tray or the platform holding the object being printed up or down (the z axis).

If you've ever looked into your desktop inkjet printer while changing a cartridge, you'll recognize many of the parts. An inkjet is a 2-D printer, which means that it works only on the x and y axes. The motor that moves the print head back and forth is just like the ones used in the 3-D printer; the inkjet uses a roller to advance the paper along the other axis. Overall, the concept is exactly the same: a computer translates a design into motor commands and deposits a material in exactly the right place, very quickly. The 3-D printer does the same thing with more motors and squirts more than just ink.

Some 3-D printers, such as the MakerBot, squeeze melted ABS plastic out a tiny hole to lay down materials in layers, a process called fused deposition modeling (FDM). Other higher-end machines use lasers to harden liquid resin in a bath (known as stereolithography, or SLA) or harden layers of powdered plastic, metal, or ceramic, a process known as selective laser sintering (SLS). The laser-driven machines can use a wider range of materials and achieve higher resolution, but tend to be more expensive than the plastic-extruding 3-D printers, which are more commonly found in homes. In a sense, this is a bit like regular paper printers, where laser printers are mostly in offices and inkjets mostly in homes.

3-D printers are an "additive" technology, which is to say that they build up objects, layer by layer, from the bottom up. By contrast, other computer-controlled machines, such as the CNC router and CNC mill, are "subtractive"; they use a spinning tool to cut or grind away material. So an additive process deposits material where the object "is"; a subtractive process takes away material where the object "isn't."

With a 3-D printer, software first examines the CAD file for an object and figures out how to make it printable using the least amount of material and time. Take, for example, a bust of a human head. The external walls of the head must be printed, but their width may be arbitrary, depending on the material used; the software will calculate the best values to print as little as possible while maintaining sufficient strength.

Typically the inside of the head is not visible, so there is no need to print it. But without any interior structure, the head could be weak and brittle. So the software will typically create a honeycomb-like support matrix inside the head, to provide the maximum amount of rigidity with the minimum amount of material (when you upload an object to a 3-D printing service bureau, you typically pay for the amount of material used or the time it takes the job to run on the machine).

The software then "slices" the object into horizontal layers as thin as the printer can handle. Each of those slices is a set of commands to the printer head to move in the x and y direction while it is extruding material or shining its laser on the powder or resin. As the head moves over the build area, it will trace out the entire slice of the object, with the software picking a path that minimizes the distance the head must move.

In a sense, this is the same concept as the original Postscript printer language that started the desktop publishing movement nearly thirty years ago. It's a way of translating from a visual language that people understand (words and typefaces in desktop publishing then, 3-D objects on a screen now) to a machine language that computers understand. Today the fabrication language is called "G-code." Just as Postscript was originally intended to drive huge industrial printers but has now found its way to the desktop, G-code was designed for machine shops but is now used in basements.

Once the 3-D printer has finished one slice, the G-code commands the z motor to move the head up a tiny fraction of an inch, and the head traces the next slice, laying down another layer of material. And so it goes, layer by layer, all the way up the object until it is finished.

In some 3-D printers, such as those that harden liquid resin, the object actually moves down into the bath as it is fabricated, so that a new layer of liquid can flow on top of the previous layer, to be hardened by the laser. This can work on a resolution as small as a few dozen nanometers, allowing the printing of structures as small as a human cell. Yet others use layers of very thin plastic sheet, with glue between each layer, and the printer head cuts out the shape in each layer. But the basic concept is always the same: build up an object in slices as thin as is physically possible. In a high-quality printer, these are practically invisible.

One of the advantages of the 3-D printer that uses a laser to harden powder is that the unhardened powder, which is still densely packed in the tray, can serve as a structural support for overhanging parts of the object, which can be droopy until they cool. When the project is done, operators take the part out and brush away the excess powder. It's possible to do the same thing with a 3-D printer that extrudes molten plastic, but ideally only with a second head, which deposits a layer of powder or other disposable material where a pillar must go to support some overhanging ledge at a higher layer.

All these manufacturing calculations sound very fiddly, but it all happens automatically; indeed, it's almost magical to watch. That's the beauty of digital fabrication; you don't need to know how the machines do their work, or how to optimize their toolpaths. Software figures all that out. The CAD design of the object contains all the information the 3-D printer needs to figure out how to make it.

The Homebrew Printing Club

This all started in industrial tooling companies in the 1980s, but over the past decade the technology has spread to regular folk, just as the PC did. To see how, take the subway to an otherwise undistinguished part of Third Avenue in Brooklyn, and knock on the metal door with the big mobile-phone readable QR code on it. Wait for some stylishly disheveled young man to open it and let you in.

Welcome to the Botcave.

In this converted brewery, Bre Pettis, Zach Smith, and their team of hardware engineers at MakerBot Industries are making the first mainstream $1,000 3-D printers. Rather than using a laser, the MakerBot Thing-O-Matic printer builds up objects by squeezing out a

0.33-mm-thick thread of melted ABS plastic, which comes in multi-colored reels.

Where industrial 3-D printers tend to look like medical equipment, MakerBots tend to be personalized, decorated with Day-Glo letters, and showered with parental love by their owners. The one I made is black with orange lettering and blue LEDs. It looks really cool when it's running in a dark room.

Out of the box, the MakerBot is a regular 3-D printer: it produces plastic parts from digital files. Want a certain gear right now? Download a design and print it out yourself. Want to modify an object you already have? Scan it, tweak the parts you want to change with the free SketchUp software from Google, and load it into the ReplicatorG app. Within minutes, you have a whole new physical object: a rip, mix, and burn of atoms.

The MakerBot is one of the simplest 3-D printers. It has just four motors: the x, y, and z, along with a fourth motor to drive the ABS plastic filament through a heater to melt it and then onto the build platform to make the object. The frame of a MakerBot is laser-cut plywood, and some of its plastic pulleys are actually made by other MakerBots themselves. The electronics are based on the Arduino processor board.

There are way more blinking LEDs than are necessary. If you have to ask why, you're missing the point. MakerBot is not just a tool. It's also a plaything. It's a revolutionary act. It's a kinetic sculpture. It's a political statement. It's thrillingly cool.

I'll bet you can't say that about your desktop inkjet.

This is the difference between commercial industrial tools and the products of the DIY movement. The Maker gear is as much about its process of creation as it is about the product itself. The fact that a MakerBot was designed by a community, manufactured by people whose names you know and whose vision you admire, and infused with personality is what makes it special. Buy one and you're not just buying a printer—you're buying a front-row seat to a cultural transformation. Open source is not just an efficient innovation method—it's a belief system as powerful as democracy or capitalism for its adherents.

The philosophy in a MakerBot goes deep. It's built on several previous open-source projects including the RepRap 3-D printer (a clever but more spindly design), the Arduino microprocessor board, and a series of software packages that turn CAD files into instructions for the four motors that control a 3-D printer. In this case, *open source* means open everything: electronics, software, physical design, documentation, even the logo. Practically everything about the MakerBot was either developed by a community or given to one to do with as they please. It is a shining example of how abandoning intellectual property protection can actually grant even more protection in the form of community support and goodwill.

I first visited the Botcave in 2009, a few months after MakerBot got started. In the long brick-walled room, one hundred boxes containing the ninth batch of MakerBots were lined up and gradually being filled up with kit parts. (As a customer, I was thrilled to know that one of them—serial number 400—was coming to me. It's gotten a lot of use since then, and I've since upgraded to a second-generation machine—the "Thing-O-Matic.") Racks of components were lined up for the next batch, and laser cutters were humming their way through stacks of thin plywood for the frames.

The creators were learning the realities of supply-chain management the hard way—those boxes couldn't go out until the last parts were in them, but some components hadn't arrived in time and others had arrived defective. A MakerBot has hundreds of parts, and if just one of them is missing, it can't ship.

The alternative to what I was seeing—scores of boxes waiting for weeks to be completed—is to over-order everything to ensure that all the components are always in stock. That's an expensive form of insurance; at the time I was there, MakerBot already had nearly $300,000 worth of parts inventory, and was still out of stock on key parts. That sort of dead capital locked up in component inventory is painful, especially for a startup. So, after focusing so hard on research and development, the team was turning to the more prosaic, but equally important, matter of securing reliable supplies of parts and forecasting demand. That's something anyone who has been in manufacturing in the past century would recognize, but it was new to this team of open-source hardware hackers. Revolutions don't come from the establishment.

As I write this, more than 5,200 MakerBots have been sold (more than $5 million worth), and with every one, the community comes up with new uses and new tools to make them even better. For example, the latest head delivers a resolution of 0.2 mm. Another head can hold a rotating cutter, turning the printer into a CNC router. Others have been scaled up to make objects twice as large as originally designed.

To date, MakerBot has raised $10 million from investors, including Amazon founder Jeff Bezos, to fund its expansion. It will need all that and more—it is competing with a host of other low-cost 3-D printer makers, including Chinese ones. What is now designed to be a kit (although you can buy it preassembled) will soon be mass-manufactured and available even more cheaply, by MakerBot and others. All the steps in using it will become easier. The market will grow from the first 5,000 to the next 50,000, from the technically sophisticated early adopters to people who just want to print something cool.

Meanwhile, the huge printer companies such as Hewlett-Packard are hovering in the wings. Right now they're just selling expensive 3-D printers to commercial customers. But at some point, probably in the next few years, the market will be ready for a mainstream 3-D printer sold in the millions in Wal-Mart and Costco. At that point, the incredible economies of scale that an HP or Epson can bring to bear will kick in. A 3-D printer will cost $99 and everyone will have one.

The gateway drug

3-D printers are appropriately mind-blowing, but while they evolve their way to eventually becoming proper matter compilers, the real workhorse of the Maker Movement is the humble laser cutter. Go into any makerspace, and the row of laser cutters are the ones working all day, with a line waiting to use them. They're the digital tool everyone uses first, in part because they're so simple and foolproof. Makers call them the "gateway drug" to digital fabrication.

Like all the digital manufacturing tools, a laser cutter is another kind of CNC machine. In this case, the computer drives motors that move a high-powered laser around an xy plane. The laser can either burn a thin line through a sheet of material (anything from plywood and plastic to thin metal) or, by varying its intensity, burn partly through it, in a form of etching.

What makes laser cutters so popular is that they're easy to use. Rather than designing an object in 3-D, you can just create an image in a 2-D drawing program such as Adobe Illustrator. Anyone can draw in 2-D—it's what we do on paper. And if you can draw it, the laser cutter can cut it out for you. It's perfect for the kind of thing you might otherwise use a jigsaw for. They're fast, cheap, and quiet—the ideal starter prototyping tool.

But just because a laser cutter works in two dimensions doesn't mean it can't make 3-D objects. Special software packages can take a 3-D object and break it out into 2-D planes, which can be cut separately, even adding little tab-and-slot elements so that they can fit together, making a strong and easy-to-assemble kit. If you've ever seen one of those wooden dinosaur skeleton kits, you've seen the work of a laser cutter.

Dozens of service bureaus, such as Ponoko or Pololu, will let you upload your 2-D file, automatically check it for errors, and help you pick an appropriate material to cut. All the parts you can fit on a plywood or plastic sheet a square foot might cost fifteen dollars. A week later your parts will arrive at your door.

If you want to cut something thicker, bigger, or less flat, you'll need a CNC router or milling machine. These are just like 3-D printers in that they operate on the x, y, and z axes, but rather than depositing material, they cut it away. Unlike a laser cutter, CNC routers can cut precise depths, too, so you can create a true 3-D object in one pass. More sophisticated "five-axis" industrial versions can twist and rotate the cutting head like a human hand to cut in from side angles and otherwise carve metal like the most skilled sculptor, but operating at superhuman speed.

I've got a desktop version called MyDIYCNC that costs $500 and uses a cheap handheld Dremel tool as its cutting head. We use it for carving desktop wargame model landscapes out of Styrofoam with the kids. We got the idea from one entrepreneur who will take your favorite video-game "map" and turn it into a scale tabletop surface under glass (the ones from the Halo series are particularly popular). It's not something we do often, but it's appropriately educational for the kids. And we can even swap a laser for the Dremel tool, and it can act like a laser cutter.

If you've recently had a cabinet maker redo your kitchen, odds are they used a bigger CNC router called a ShopBot. If you've bought flat-packed IKEA furniture, that was CNC'd at the factory. Your car was probably prototyped with a room-sized CNC machine, which carved the body shape out of foam. And even bigger, warehouse-sized CNC machines can carve an entire airplane fuselage out of foam, which will serve as a mold for a fiberglass body.

Reality capture

All these digital tools are ways to turn bits into atoms. But how about the reverse: turning atoms into bits? It's hard to draw 3-D objects from scratch on a screen; much easier is to just start with something that already exists and is similar to what you want, and then modify it.

This process is called "reality capture." The idea is that you can take any object and scan it, creating a "point cloud" of dots that define its surface. Then other software turns that cloud of points into a mesh of polygons, just like the "wireframes" that make up the characters in computer-animated films, which can be manipulated and modified onscreen.

You can buy a commercial 3-D scanner that can do this with lasers that trace over an object and cameras that capture the positions of points on its surface, but there are cheaper ways, too. Autodesk offers a free online service called 123D Catch that allows you to upload regular photographs of an object (taken from all angles), and cloud-based software will turn it into a 3-D object that you can modify and print on a 3-D printer. There's even a version that runs on the iPad.

Or you can make your own 3-D scanner with a pocket projector shining a grid pattern ("structured light") on an object, which is viewed with a high-definition webcam. Rotate the object and a special webcam program will capture all the sides and dimensions, extracting geometries from the way that a known light pattern is distorted when projected on the surface of the object.

Finally, there are research projects to do this with the webcam built right into your laptop or smartphone. Software running on your PC can guide you into rotating and showing different sides of the object, filling in the missing pieces in the software's internal model of it. This sort of "guided scanning" can mean that someday if you want to duplicate an object, you need only point your phone at it, following the phone's directions to move around the object and zoom in on sections, and press "print." A duplicate, perhaps even in color, will appear in your desktop 3-D printer.

At that point the possibilities become clear. We can photocopy reality, or at least as closely as a Hollywood prop. And the resolution will only improve. Low fidelity will become high fidelity. The next step will be to go more than skin-deep, duplicating not just form but also function. We can already make the cup for the Earl Grey tea. How much longer until we can make the tea, too?

The replicator awaits.

Chapter 7

Open Hardware

A market where customers help you develop your products and then pay you for them? Sure— just give away the bits and sell the atoms.

One sunny Friday afternoon in March 2007, I started planning what I'd hoped would be a delightfully geeky weekend with the kids. In the usual stack of boxes that had come into the *Wired* offices that day to be reviewed, there was a Lego Mindstorms robotics kit and a ready-to-fly radio-controlled airplane. I claimed them both, promising to write the reviews, and settled on a schedule: we would build robots on Saturday and fly planes on Sunday. Awesomeness surely awaited.

But by midmorning on Saturday, things were already going wrong. The kids were happy enough to open the Lego Mindstorms box and assemble the starter robot, a three-wheeled rover, but once we plugged in the batteries they could barely hide their disappointment. Hollywood, it turns out, has ruined robotics for kids: they expect laser-armed humanoid machines that can transform into trucks. Meanwhile, after an hour of assembly and programming, the Mindstorms rover could only roll forward and bounce feebly off a wall. We looked online to see what others were doing with Mindstorms, and saw that hobbyists had already made everything from robotic Rubik's Cube solvers to working photocopiers. We wanted to invent something new, but there was no way we could do that sort of thing, or anything even close to it. The kids lost interest after lunch.

Okay, there was always the plane. On Sunday we took it to a park. I tossed it in the air and promptly flew it into a tree. The kids just looked at me, equally appalled by my lack of ability and the gap between my promise of how cool the plane would be (and the spectacular YouTube videos of aerobatics we'd watched) and how uncool it had actually turned out to be. I threw sticks at the plane in the tree to try to dislodge it, while my mortified children pretended not to be with me. My geekdad weekend was a failure, and I was equally annoyed at myself for getting it so wrong and at my kids for being so unappreciative. I went for a run to let off some steam.

While on the run, I started thinking more about the sensors that were available for Lego Mindstorms. There were accelerometers ("tilt sensors"), electronic gyroscope sensors, a compass sensor, and a Bluetooth link that could connect to a wireless GPS sensor. They were actually pretty amazing, and it occurred to me that those were exactly the same sensors that you'd need to make an airplane autopilot. We could kill two birds with one stone: invent something cool with Lego that had never been done before and get the robot to fly the plane! It was sure to be a better pilot than me.

The moment I got home, I prototyped a Lego autopilot on the dining room table, and my nine-year-old helped write the software. We took some pictures, posted them, and it was on the front page of Slashdot by that evening. We put it in a plane—the world's first Lego UAV, I think—and took it out a few weekends later. It almost kinda worked—it was definitely staying aloft and steering on its own, albeit not exactly where we had intended.

At that point I went down the rabbit hole and resolved to improve it until it worked as I'd dreamed, a quest that I'm still on years later. (The kids, sadly, lost interest within days, and returned to their usual staple of video games and YouTube, both of which offer more immediate gratification.)

I worked on some improved versions of the Lego autopilot, eventually developing one that had most of the mission-following functionality of a professional autopilot, if not the performance (it's now in the official Lego museum in Billund, Denmark). But it was soon clear that Lego Mindstorms, for all its charms, was not the right way to make a real autopilot: it was too big and expensive, for starters, and didn't then have a good way to work with radio-control systems.

What would be a better way? I decided to conduct my search for answers online in public, sharing what I'd done and found. But because this was 2007 and Facebook was booming, I set up DIYDrones .com as a social network (on the Ning platform), not as a blog (so 2004!).

That distinction—a site created as a community, not a one-man news and information site like a blog—turned out to make all the difference. Like all good social networks, every participant, not just the creator, has access to the full range of authoring tools: along with the usual commenting, they can compose their own blog posts, start discussions, upload videos and pictures, and create profile pages and send messages to one another. Community members can be made moderators, to encourage good behavior and discourage bad.

What this meant was that the site wasn't just about me or my ideas. Instead, it was about anyone who chose to participate. And right from the start, that was almost everyone. The site was soon full of people trading ideas and reports of their own projects and research. Initially, members would just post code and design files, trading ideas back and forth in a form of nerd braggadocio. But over time we set up more organized systems of collaboration, including version control systems and file repositories, wikis, mail lists, and formal team assignments.

I was blown away by what I was seeing people in our community doing with sensors from mobile phones and chips that cost less than a cup of coffee; feature-by-feature they were equaling aerospace electronics that had cost millions of dollars just a decade earlier. It felt like the future of aviation: just as the PC emerged from the Homebrew Computer Club hobbyists and eventually overturned the industrial corporate computing world in the 1980s, I could imagine that the same sort of movement could be the way that robots were introduced to our skies. We were present at the creation. If there was going to be an Apple Computer of this industry, it should be us!

At this point my entrepreneur instincts kicked in. Something in my wiring forbids the notion of fun for its own sake; instead, everything must be building to a purpose. What this normally means is an unfortunate tendency to "industrialize my hobbies," which usually has the sad effect of making them not fun anymore. (I had done the same a few years earlier with parenting. My search for fun technology projects to do with the kids turned into the GeekDad blog, which is now GeekDad Inc., a successful stand-alone company. At least in that case, I was able to turn it over to others before parenting started to feel like a job.) It was soon clear that DIY Drones would be no exception.

My first cottage industry

I started my first aerial robot business on the proverbial dining room table. Using a blimp controller design created with a community member, I started assembling the parts necessary for a kit. I sent the circuit-board design files off to be fabricated, and began hunting around for good deals on other electronics parts that I could buy in volume. Weeks of sourcing followed, with the simple rule that manufacturers should never pay retail for their materials. Motors from China, Mylar blimp "envelopes" from a warehouse in Canada, propellers from Taiwan, a big box of custom laser-cut plastic shapes for the base, and stacks of flat-packed cardboard pizza boxes to put them in. (I also got Lego to donate a big box of gears and shafts.)

The first few dozen boards I hand-soldered together, before vowing to Never Do That Again. Then I advertised on Craigslist for a local student to do another hundred or so, which turned out to be more trouble than it was worth. Finally I just did what I should have done from the start and contracted with an assembly firm to do a few hundred more properly, with automated pick-and-place machines. A big box of the finished boards arrived at my doorstep, and I spent an evening testing them and loading their software.

Finally it was time to pack the kits. We had all the components, and I bribed the kids to be my packing team. Piles of parts with Post-it notes specifying how many should go in each box spread over the dining room table and floor. For a full morning, as they filled box after box with increasing tedium, the kids knew what it was like to work in a real factory. (A painful lesson: don't put a five-year-old on quality assurance—we had to check all those boxes again.)

For the community's next product, an airplane autopilot board, we decided to put things in the hands of professionals. The one that seemed culturally matched was Sparkfun, which designs, makes, and sells electronics for the growing open-source hardware community. Because they handled all the sourcing and manufacturing, our community could spend its time working on R&D and bear no inventory risk.

But, over time, our community started designing products faster than Sparkfun could adopt them, and many were too niche for Sparkfun's store. It was time to start our own factory. I started a proper company, 3D Robotics, with a partner, Jordi Muñoz (of whom much more later).

In a rented Los Angeles garage, Muñoz started building our own mini Sparkfun. Rather than a pick-and-place robot, we had a kid with sharp eyes and a steady hand, and for a reflow oven we used what was basically a modified toaster oven. We could do scores of boards per day this way.

As demand picked up, we outgrew the garage. Muñoz moved the operation to commercial space in an industrial park in San Diego, which was nearer the low-cost labor center of Tijuana. In came real automated manufacturing tools: first a small pick-and-place machine, then a bigger one, and finally an even bigger one with automated component feeders. The toaster oven gave way to a proper automated reflow oven with a nitrogen cooling system for perfect temperature control. And for that we needed a nitrogen generator, of course. And so it went, with more and more professional tools, which Muñoz and his team learned to use by finding tutorials on the Web.

By this time we had outgrown the first space and expanded to a bigger space next door. Then we outgrew that, too, and today 3D Robotics has a factory that sprawls over twelve thousand square feet and a second one of nearly the same size in Tijuana. The facilities are buzzing with robotic assembly machines run by factory workers, and teams of engineers developing new products. Pick-and-place robots build circuit boards, which are baked in automated reflow ovens temperature-regulated by a nitrogen generator. Laser cutters, 3-D printers, and CNC machines make quadcopter parts. These are real factories now, just three years after Muñoz started hand-assembling boards on his kitchen table with a soldering iron.

From Maker to millions

In our first year, we did about $250,000 in revenue; by 2011, our third year, we had broken $3 million. In 2012 we're on track to break $5 million in revenues. Growth continues at about 75 to 100 percent per year, which is common for open-source hardware companies like ours. We've been profitable from the first year (it's actually not that hard in the hardware business—just charge more than your costs!), but try to reinvest as much of the profits as possible into building new factory lines. Because we're online, we're global from the start and tend to grow more quickly than traditional manufacturing companies because of the network effects of online word of mouth. But because we're making hardware, which costs money and takes time to make, we don't show the hockey-stick exponential growth curve of the hottest Web companies.

So, as a business, we're a hybrid: the simple business model and cash-flow advantages of traditional manufacturing, with the marketing and reach advantages of a Web company. We're still a small business, but the difference between our kind of small business and the dry cleaners and corner shops that make up the majority of micro-enterprise in the country is that we're Web-centric and global.

We're competing in the international market from day one. The usual trap of focusing on the local market first with hopes of expanding internationally later leaves companies unprepared for global competition. Selling to the whole world on day one makes a company stronger. Today, two-thirds of our sales come from outside the United States. And with global reach comes the capacity to grow far beyond what local markets could support.

Make a profit. It's not that hard.

Profit is always a tricky question for Web companies, since they tend to put a priority on growing traffic, and charging money gets in the way of that. But for hardware, which has inherent costs and must be paid for, charging the right price is key to building a sustainable business.

One of the first mistakes budding Makers make when they start to sell their product is not charging enough. It's easy to see why, for all sorts of reasons. They want the product to be popular, and they know the lower the price, the more it will sell. Some may even feel that if the product was created with community volunteer help, it would be unseemly to charge more than it costs.

Such thinking may be understandable, but it's wrong. Making a reasonable profit is the only way to build a sustainable business. Let me give you an example. You make one hundred units of your delightful laser-cut handcrank toy Drummer Boy. Between the wood, the laser cutting, the hardware, the box, and the instructions, it costs you $20 to make each one. Let's say you price them at $25 just to cover any costs you may have missed, and start selling.

Since it's a fun kit and pretty cheap, it sells quickly. You suddenly realize that you've got to do it all again, this time in a batch of one thousand. Rather than putting up a couple thousand dollars to buy the materials, you've got to put up tens of thousands of dollars. Instead of packing the kits in your spare time, you've got to hire someone to do it. You need to rent space to store all the boxes, and you've got to make daily trips to FedEx.

Now your hobby is starting to feel like a real job. Worse, the popularity of your kit has come to the attention of some big online retailers, and they're asking about buying in batches of one hundred, with a volume wholesale discount. You're thrilled that your kit is so popular and flattered that these retailers, who can reach many more people than your own website, want to sell it. But if you're selling it at $25, that's the market price—the retailers typically can't sell it for more. The retailers ask for a lower price because they need to make their own profit on each one, usually around 50 percent. So they need to buy them at no more than $17 each. But that would mean you are selling each one at a loss! Your costs, which were once within the limits of hobby spending, are now at risk of driving you and your business into debt.

What entrepreneurs quickly learn is that they need to price their product at least 2.3 times its cost to allow for at least one 50 percent margin for them and another 50 percent margin for their retailers (1.5 × 1.5 = 2.25). That first 50 percent margin for the entrepreneur is really mostly covering the hidden costs of doing business at a scale that they hadn't thought of when they first started, from the employees that they didn't think they'd have to hire to the insurance they didn't think they'd need to take out and the customer support and returns they never expected. And the 50 percent margin for the third-party retailers is just the way the retail market works. *(Most companies actually base their model on a 60 percent margin, which would lead to a 2.6x multiplier, but I'm applying a bit of a discount to capture that initial Maker altruism and growth accelerant.)*

In other words, that $20 kit should have been priced at $46, not $25. It may sound steep to you now, but if businesses don't get the price right at the start, they won't be able to keep making their products, and everyone loses. It's the difference between a hobby and a real, thriving, profitable business. It's also worth bearing in mind that at this more bespoke end of the market, products can generally support a higher price. Customers are both keen and savvy: they are prepared to spend a bit more because they know that they are getting exactly what they want. It's an attractive business model.

Open design's advantage

Today we use the products of open *software* innovation every day: the Firefox Web browser, Android phones, the Linux Web servers that run most of the websites we go to, and countless other elements of the open-source software that the Internet is built on. Tomorrow the same may be true for *hardware*, too. I've driven in open-source cars (the Local Motors Rally Fighter, of which you'll hear more later) and watched open-source planes fly. There are open-source rockets designed to reach space, as well as open-source submarines. We have open-source watches and alarm clocks, cappuccino makers, and toaster ovens.

In a sense, all these companies *give away the bits and sell the atoms*. All the design files, software, and other elements that can be described in digital form—the bits—are given away freely online, under a license that usually offers almost unrestricted use as long as it continues to be open and shared. But the physical products themselves—the atoms—are sold, because they have real costs that must be recouped.

Every day, we see more and more examples of open-hardware business models working brilliantly. The MakerBot 3-D printer is open hardware, as is the RepRap on which it is built. So is Arduino, and the hundreds of products from companies such as Adafruit, Seeed Studio, and Sparkfun. Research by Adafruit's Phillip Torrone concludes that there were more than three hundred commercial open-hardware products by the end of 2011, representing more than $50 million in annual revenues.[28]

Openness, in fact, is exactly what Thomas Jefferson and the Founding Fathers intended when they made the Patent Act one of their first orders of business in the new United States of America in 1790, a year after the Constitution was ratified. As they saw it, the point of a patent—a guaranteed monopoly granted for a limited time—was not primarily to ensure that the inventor made money; after all, they could do that more easily by keeping the invention a trade secret. Instead, it was to encourage the inventor to *share* that invention publicly so that others could learn from it. The only way an inventor could license a patent was if he or she published it, ensuring that society as a whole could benefit from the invention. (Science works the same way, with credit and career advancement depending on publication in journals.)

Today, inventors increasingly share their innovations publicly without any patent protection at all. That is what open source, Creative Commons, and all the other alternatives to traditional intellectual property protection do. Why do they do so? Because the creators believe they get back more in return than they give away: free help in developing their inventions. People tend to join promising open projects, and when those projects are shared, the contributions are automatically shared, too. Inventors also get feedback as well as help in promotion, marketing, and fixing bugs. And they accrue "social capital," a combination of attention and reputation (goodwill) that can be used at a future date to advance the inventor's interests.

A product that has successively been created in an open innovation environment does not have the same legal protections as a patented invention. But one can argue it has a better chance of becoming a commercial success. Odds are that it was invented faster, better, and more cheaply than it would have been if it had been created in secret. It's already been tested in the marketplace of opinion, at least, and that's not a bad form of market research. And it's got a built-in marketing team in its community, evangelists who are invested in its success. Any product that can build a community before launch has already proven itself in a way that few patents can match.

For the companies that are built on open innovation, the advantages go beyond simple access to markets. A well-constructed "architecture of participation," to use the term coined by Tim O'Reilly, whose company runs *Make* magazine, means that hundreds of skilled people may contribute for free, for all the incentives that have been observed in everything from open-source software to Wikipedia, from being part of something they believe in to simply making something that serves their own needs, but choosing to share it because of the community norms.

What that means is cheaper, faster, and better research and development, which in turn can create unbeatable economics for companies whose products are developed this way. And it's not just R&D. Product documentation, marketing, and support are often done the same way, by a community of volunteers within a community. Some of the most costly functions of traditional companies can be done for free, so long as the social incentives are tuned right.

It's what we do for everything at 3D Robotics, and here's why: when you release your designs on the Web, licensed so that others can use them, you build trust, community, and potentially a source of free development advice and labor. We release our electronics PCB designs in their native form (Cadsofts' Eagle format), under a Creative Commons Attribution ShareAlike license ("by-sa"), which allows commercial reuse. Our software and firmware, meanwhile, are all released under a GPL license, which also allows for commercial reuse as long as attribution is maintained and the code stays open. The result: hundreds of people have now contributed code, bug fixes, and design ideas, and have made complementary products to enhance our own.

The simple act of going open source has provided us with an essentially free R&D operation that would have cost hundreds of thousands of dollars if we'd been closed-source and had to hire our own engineers to do the work, to say nothing of the quality of that work. By day our volunteers are leading professionals in their own fields—the sort that would have been impossible to hire away. But by night they follow their passions and do great work for us as volunteers. They do it because we're collectively making something they both want for themselves and want to be part of, and because it's open source they know that it will reach more people and attract more talent, creating a virtuous cycle that accelerates the innovation process far faster than conventional development can.

Once you seed your community with content and start attracting users, your job is to give *them* jobs. Elevate people who seem to be constructive participants to moderator status, and give especially friendly and helpful members a "noob ninja" badge. Once you promote/reward enough of them for doing a good job of constructive community building, you'll find that members typically help one another, saving you the work.

Finally, the tricky matter of whether to pay volunteers: I'm in favor of offering key contributors to a product a royalty, but don't be surprised if they decline. The reasons can be many: they're not in it for the money; the absolute payment amounts are tiny compared to what they make in their day jobs; they feel it's wrong taking payment when others who contributed don't; and finally, when they realize that any royalty you pay will lead to higher prices for consumers, they decline simply because this conflicts with the real reason they contributed, which is to create something that can reach the largest audience possible, and higher prices mean fewer users.

But there are rewards short of simple payment that can be even more motivating, especially for top contributors, who tend to be equally accomplished in their professional lives.

Here, for example, is the reward hierarchy we came up with for the DIY Drones dev teams. It ranges from the silly but effective, such as a coffee mug for a "commit" (a commit is a code contribution of any size, which may have taken only an hour or two), to a reward that could have significant monetary value, such as stock options in 3D Robotics for top contributors.

The Hierarchy of Reward

Team leaders who have shipped a major project — EQUITY

Core Team leaders — TRIP TO DEV MEETING

Accepts role as a project leader — DEV TEAM MEMBER AND FREE HARDWARE

"Sustained Contribution Award" by team leader — COFFEE MUG AND HARDWARE DISCOUNT

First accepted commit — T-SHIRT

How to build community

When you go open source, you're giving away something in hopes of getting back more in return. Is it guaranteed? No. You also need to build a community, ensure that the initial product is needed, documented, and distinct enough for people to want to join in its development. And even then, managing an open-source community can be a full-time job in itself. But when it works, it can be magical: an R&D model that's faster, better, and cheaper than those of some of the biggest companies in the world.

When you're creating a community from scratch, consider starting it as a social network rather than as a blog or discussion group. The best new social networking tools allow you to have it all: great blogging tools, great discussion groups, profiles, personal messaging, videos, photos, and more.

One of the key elements of a successful community is content with broad appeal, not just discussion forums but blog posts, photo and video sharing, and news feeds. The Makers communities all have this, from the fantastic stream of daily blog posts of MakerBot, Sparkfun, and Adafruit to the video profiles of members on Kickstarter and Etsy.

In a sense, such rich, engaging content is marketing—marketing for the community itself, but also for the products that the community has created. Whether they think of it this way or not, the most successful Makers are also the best marketers. They're constantly blogging about their progress, and tweeting, too. They take pictures and videos of every milestone, and post those. Their excitement in making is infectious, and builds excitement and anticipation for the products they ultimately release.

Seen this way, all making in public is marketing. Community management is marketing. Tutorial posts are marketing. Facebook updates are marketing. E-mailing other Makers in related fields is marketing. Of course, it's not *just* marketing: the reason that it's so effective is that it's also providing something of value that people appreciate and pay attention to. But at the end of the day, everything you do, from the naming of your product to whose coattail you decide to ride (like we chose Arduino), is at least partly a marketing decision. Above all, your community is your best marketing channel. Not only is that the source for the word-of-mouth and viral marketing that you'll need, but it's also a safe place to talk about your own products, as enthusiastically as you want. If you've given people a reason to gather that serves their needs and interests, crowing about your cool new gizmo isn't advertising, it's content!

Castles without walls

But how can companies built on such open-innovation grounds protect themselves from competition and even piracy? After all, part of the social compact in open innovation is to return the gift, sharing everything with the community that created it. What is their defensible advantage?

Is it the brands? Many open-hardware projects share the design files of their products, but reserve their names and logos as proprietary trademarks. Others can make the same product, but they can't *call it the same product* (at least not legally in the countries where the trademarks are registered). Brand can indeed be a defensible advantage. But the legal process of fighting infringement of those brands, especially in other countries, can be ruinously expensive. And in an open-source world you can't assume the clones will be inferior and easy to spot.

Is it the communities? Yes, as long as they're serving early adopters and other Makers. A Chinese company can make a clone of our products and maybe sell it cheaper, but it won't have our community, and if our community can spot the clone, they will probably decline to help those who chose not to support the "home team." But let's be honest: our communities exist because our products are *hard to use*. They are mostly support communities, where members help each other navigate confusing and uncharted territory. There's also a bit of a development community, for the one percent of users who also want to help evolve the products or take them in new directions.

Ultimately, however, the goal of open-innovation projects is to make products that are as good as or better than traditional closed-innovation products. And that means easy to use: well designed and documented. When you go to Crate & Barrel to buy a toaster, you don't care if it has a community. Great products don't need great communities. Sometimes great products speak for themselves.

In those cases, the only real defensible advantage is an ecosystem. Not a community of customers, but a community of other companies and innovators who are building products that are designed to work with and support your own. Think of the tens of thousands of apps that support and reinforce Android, an open (mostly) mobile operating system. Or the hundreds of plug-ins and utilities designed to work with WordPress, the open-source blogging platform. In each case, openness built a constituency for the product's continued success. The fact that others could copy it didn't matter, because all that goodwill had created a network effect that was far harder to copy than simply code.

But what if someone wants to rip us off anyway? Well, it depends on what you mean by "rip us off." If someone else decides to use our files, make no significant modifications or improvements, and just manufacture them and compete with us, they'll have do so much more cheaply than we can to get traction in the marketplace. If they can do so, at the same or better quality, then that's great: the consumer wins and we can stop making that product and focus on those that add more value (we don't want to be in the commodity manufacturing business).

But the reality is that this is unlikely. Our products are already very cheap, and the robots we use for manufacturing are the same ones they use in China, at the same price. There is little labor arbitrage opportunity here.

And even if the products can be made cheaper, at the same quality, there is the small matter of customer support. Our community is our competitive advantage: they provide most of the customer support, in the form of discussion forums and blog tutorials and our wiki. If you bought your board from a Chinese cloner on eBay and it's not working, the community is unlikely to help—it's seen as not supporting the team that created the product in the first place.

How will people know the difference between our products and clones allowed by our open-source license? Because the clones can't use the same name. The only intellectual property that we protect is our trademarks, so if people want to make the same boards, they'll have to call them something else. This is the same model used by the Arduino project. You can make a copycat board, but you can't call it "Arduino" (although you can call it "Arduino compatible"). This goes all the way to removing the logo, name, and artwork from the PCB design files that are publicly distributed. It's a great way to maintain some commercial control while still being committed to the core principles of open source.

Another core aspect of open source is that users can make the products themselves, if they want—no need to pay you for it. That's great for about 0.1 percent of the user base, and they're often the best source of new ideas and innovation around the product. But the reality is that the other 99.9 percent of users would rather pay someone to do it for them, guaranteeing that it will work. That's the core of your business.

How to get your "pirates" to work for you

Here's an example of how it works in practice: In late 2010, someone posted on the DIY Drones site that Chinese copies of our ArduPilot Mega design were for sale on Taobao, eBay, and other online marketplaces. And indeed they were: well-produced, fully functional clones. Not only that, but our English instruction manual had been translated into Chinese, too, along with some of the software.

Our community members were shocked by this blatant "piracy" and asked what we were going to do about it.

Nothing, I said.

This is both expected and encouraged in open-source hardware. Software, which costs nothing to distribute, is free. Hardware, which is expensive to make, is priced at the minimum necessary to ensure the healthy growth of a sustainable business to ensure quality, support, and availability of the products, but the designs are given away free, too. All intellectual property is open, so the community can use it, improve it, make their own variants, etc.

The possibility that others would clone the products is built into the model. It's specifically allowed by our open-source license. Ideally, people would change/improve the products ("derivative designs") to address market needs that they perceive and we have not addressed.

That's the sort of innovation that open source is designed to promote. But if they only clone the products and sell them at lower prices, that's okay, too. The marketplace will decide.

By the way, the Arduino development boards have gone through exactly the same situation, with many Chinese cloners. The clones were sometimes of lower quality, but even when they were good, most people continued to support the official Arduino products and the developers who created them. Today clones have a small share of the market, mostly in very price-sensitive markets such as China. And frankly, being able to reach a lower-price market is a form of innovation, too, and that is no bad thing.

Personally, I'm delighted to see this development, for four reasons:

1. I think it's great that people have translated the wiki into Chinese, which makes it accessible to more people.

2. It's a sign of success—you get cloned only if you're making something people want.

3. Competition is good.

4. What starts as clones may eventually become real innovation and improvements. Remember that our license requires that any derivative designs must also be open-source. Think how great it would be if a Chinese team created a better design than ours. Then we could turn the tables and produce their design, translating the documentation into English and making them available to a market outside China. Everybody wins!

Shortly after I wrote this, a member named "Hazy" responded in the comments that he had been working with the team that had made the Chinese boards, and was the one doing the translation. I complimented him on the speed at which it had been done, and then asked him if he'd consider bringing the translation into our official manual, which takes the form of a wiki on Google Code, where our repository is. He agreed to do so, and so I gave him edit permission to the wiki and otherwise set it up for a parallel Chinese translation that users could select.

At the time, we were using the Subversion version-control system (we're now using Git), and Google Code had a relatively basic implementation of it. The wiki pages were just files in the same repository as the source code for our autopilots, and I hadn't investigated the permissions options very well. To let people edit the wiki, I just gave them blanket "commit" access (the ability to create and edit files) to the whole repository.

When I gave community members such access, I usually asked them not to mess with the code by mistake (membership in the code development teams was more exclusive, because the danger of messing things up was higher), but in the case of Hazy I forgot.

The first thing Hazy did was integrate the Chinese translation of the manual seamlessly, so users could simply click a link to switch easily between the two languages.

Then, because he was an expert in our autopilot (he had, after all, been part of the team that cloned it), he started making corrections in the English manual as well. I could see all the commits flowing by and approved them all: they were smart, correct, and written in perfect English.

Then it got interesting: Hazy started fixing bugs in the code itself. The first time this happened, I assumed he'd made a mistake and pushed a wiki file into the wrong folder. But I checked it out, and it was code and his fix was not only correct, but properly documented. Who knew that Hazy was a programmer, too?!

I thanked him for the fix, and thought little more of it. But then the code commits kept coming. Hazy was working his way through our Issues list, picking off bugs one after another that the dev team had been too busy to handle themselves.

Today he is one of our best dev team members. I've still never met him, but after a while I asked him a bit about himself.

His real name is Xiaojiang Huang. He lives in Beijing, and by day he is a Ph.D. student in computer science at Peking University.

He told me his story:

When I was a kid, I was fascinated by all kinds of models, and I wished I could have an RC plane. Several years later, I was able to afford an RC helicopter when I graduated from college. I also got RC trucks and planes. Sometimes I am derided as naïve for playing with "toys," but I'm happy because it's my childhood dream. I met ArduPilot by chance when I was surfing the Web, and was attracted by its powerful functions. Some friends of mine were also interested in it, but they felt it was a little inconvenient because of the English documents. So I tried to translate them into Chinese, hoping to reduce the difficulty of playing with ArduPilot for the Chinese fans. Thank you for the great work of DIY Drones, and I hope it will help more people make their dreams come true.

What happened there is magical. When we first got word of the cloned boards, some in our community initially jumped to the conclusion that this was another case of blatant Chinese piracy, and they wanted to know when we were going to sue. But when I reminded them that this was not a "pirated" version, but instead a "derivative design" fully permitted and even encouraged by our open-source license, the tenor changed.

Because we did not demonize the Chinese team, but instead treated them as part of the community, they acted that way, too. Hazy stepped forward and, rather than just exploiting our work, contributed to it as well.

So now at least some of the "pirates" work for us. Instead of just using our technology, they're helping us to improve the technology for everyone. "Hazy" realized his dreams, and in doing so helped us realize ours, too.

Chapter 8

Reinventing the Biggest Factories of All

There's no manufacturing business like the car business. If that can be transformed, anything can.

There is no law that says that Maker companies have to remain small. After all, many of today's biggest Silicon Valley giants, from Hewlett-Packard to Apple, started in a garage, and on the Web the dorm-room-to-riches story is now so common that computer science students who stick around long enough to finish their degree risk being considered lacking in entrepreneurial gumption. As a hybrid between traditional manufacturing companies and Web startups, Maker companies also have the potential to be the next big thing, combining the growth rates of software with the money-making ability of hardware.

But at the end of the day, the Maker Movement will be judged not just on how it can change product categories and entrepreneurial fortunes, but also on how much it can move the needle for an entire economy. And to do that, it will have to be able to influence the biggest manufacturing industries—the car industry being the biggest of all. Even here, in one of the toughest of all manufacturing sectors, it's already possible to see a future for Makers. While they may not have massive economies of scale, they do have the flexibility and focus that define companies that are most connected to their customers today.

There have, of course, always been niche car companies and small suppliers in the automotive industry. But that's been an increasingly tenuous place to be, as anyone knows who has watched the gradual decline and sale of most of Britain's specialist car companies to multi-national giants. The problem is that the conventional car industry has historically proven a hostile place for innovations. To see how, consider the story of the creation of the intermittent windshield wiper.

The trials of a 20th-century inventor

On his wedding night in 1953, a young engineer named Robert Kearns was hit in the left eye with a champagne cork, rendering him legally blind in that eye. A decade later he was teaching at Wayne State College in Detroit, and among the many things that bothered him about his diminished eyesight, the distraction of the windshield wipers of his Ford Galaxie in the rain seemed like one thing he could do something about.

The wipers were annoying, and not just for people with only one good eye. When they were on, they constantly moved back and forth, regardless of how hard the rain was falling. You could slow them, but you couldn't pause them, even if it was only sprinkling. It was as if your eyelids were continually opening and closing, rather than blinking every now and then. For a man with impaired vision, the constant motion was yet another distraction while driving. For an engineering professor, it was a challenge to find a better way.

Kearns went into his basement workshop and began tinkering. He prototyped an electrical delay circuit on his workbench, which gradually charged a capacitor to pause a set of wipers for an adjustable length of time, depending on how hard it was raining. As depicted by Greg Kinnear in the 2008 movie *Flash of Genius*, which is based on this story, Kearns exuberantly demonstrates the working model to his kids with wiper motors swiped from his wife's car, and a plate of glass: "It's aliiiive!" The kids are suitably impressed and even help with soldering (this is the Hollywood version, after all). It is a pure scene of invention: one man, an idea, and the tools and skills necessary to make it real.

That picture is remarkably similar to my grandfather's, as I remember it, minus the theatrics. The difference is that Kearns then made a decision my grandfather did not. Although both filed a patent application on their invention, Kearns decided not to license his invention to the car companies. Instead, he decided to make his intermittent windshield wiper himself and sell it to them as a finished product. That meant he needed to build a factory.

Kearns borrowed money, took an investment from a partner, remortgaged his house, and otherwise scrabbled together the huge sums necessary to make a wiper factory in the mid-1960s. It was a staggering undertaking and, as events would soon show, unwise.

The scenes where he sets up the factory are telling. First, there is the renting of a 30,000-square-foot industrial space, all open areas with only columns standing between the exterior brick walls and loading docks. Then the space must be filled with production equipment. Men in hard hats carry steel racks around and drive forklifts, carrying roller bearings for conveyer belts—a classic industrial-age picture. Finally, there is the meeting with Motorola to arrange for the purchase of transistors, which requires negotiated credit from the company's finance department. Scary stuff for a small entrepreneur.

It would get scarier. As the film tells it, Kearns is close to firing up his facility when suddenly Ford stops returning his phone calls. He has no idea why, but clearly a sale to Ford isn't going to happen. With no revenue in sight, the factory shuts down before producing a single wiper.

Eighteen months later, Kearns is returning to his car in the rain and sees a trio of brand-new Ford Mustangs turn the corner, driving to their big rollout party. Their windshield wipers sweep, then pause, then sweep again. His brilliant idea, he believes, has been stolen. Kearns is ruined and will soon go mad, thus the dramatic rest of the movie. (In reality, Ford introduced intermittent wipers a few years later than the movie depicts, as an option on the 1969 Mercury line.[29] But Kearns's years of despair, depression, and breakdowns are sadly accurate.)

Kearns eventually sued Ford and Chrysler for patent infringement and, after years of litigation, Ford eventually settled by paying nearly $10 million, and Kearns was awarded $30 million from Chrysler, after $10 million of legal bills. But his fight against the car companies was not about the money, he insisted—it was about the principle of the thing. His obituary in 2005 reported that "all he wanted, he often said, was the chance to run a factory with his six children and build his wiper motors."[30] He never got that chance. It was just too hard back then.

Today Kearns would do it differently. As before, he would have made the first prototype in his basement. But rather than building a factory, he would have had the electronics fabbed by one company and the enclosure made by another. He then would have paid a wiper manufacturer in Guangdong or Ohio or any of countless other places to create a custom assembly with these components. They would probably be shipped straight to his customers, the car companies, and the whole process would have happened in months, not years—too fast for big companies to beat him. No factory, no lawsuits, no madness. He could have fulfilled his dream of turning his invention into a company without tilting at windmills.

Genius, reflashed

There's no need to imagine this scene. You can see something like it today. Just go to Chandler, Arizona, and find the Local Motors factory in a converted recreational vehicle warehouse twenty minutes south of Phoenix. Columns draped with potted plants soften the interior, a design detail borrowed from a Ferrari facility (although they're a challenge to keep healthy under artificial light), but otherwise this looks more like a car dealership than a factory; there is, for starters, no production line. Instead, individual cars are being lovingly worked on next to color-coordinated tool cabinets.

This is where the world's first open-source cars are being produced, starting with a $75,000 Baja racer called the Rally Fighter, with curves inspired by a fighter plane. The Chandler site is just the first "microfactory" of many the company plans to build across America, each with about forty employees. Each will manufacture cars created by the community, which helps build them. It's a glimpse into a whole new way to design, engineer, and produce cars—and maybe lots of other things, too.

Local Motors is a car company built on Maker principles. Its designs are crowdsourced, as is the selection of mostly off-the-shelf components. It doesn't patent ideas—the point is to give them away so that others can build on them and make them even better, for the benefit of all. It holds almost no inventory, and purchases components and prepares kits only after buyers have made a down payment and reserved a build date.

It started with a question: How would you build a car company on the Web? In 2007, Jay Rogers and Jeff Jones decided to find out. They created a site where car designers (professionals, amateurs, and those just interested in the process) could share ideas and vote on their favorites. They called the company Local Motors because they hoped that someday its manufacturing could be as geographically distributed as its community, with local "microfactories" serving as their dealerships. Rather than having a big central factory, the cars would be built on-demand by their customers, near where they live.

Rogers was practically destined for his job. His grandfather Ralph Rogers bought the Indian Motorcycle Company in 1945. When the light Triumph motorcycles began entering the United States after World War II, the senior Rogers recognized that his market-leading Chief, a big road workhorse, was uncompetitive. The solution was to make a new light engine so Indian could produce its own cheap, nimble bikes. He went bust trying to develop the motor. It was just too hard and expensive to change direction—and eventually he lost the business.

Today, Ralph Rogers's grandson intends to do something even more radical—create a whole new way of making cars—on a shoestring budget. It's just easier these days. His company has raised roughly $10 million, and he thinks that's enough to take it to profitability.

The difference between now and then? "They didn't have resources back then to enter the market, because the manufacturing process was so tightly held," he says. What's changed is that the supply chain is opening to the little guys.

Rogers and Jones believed that open innovation could change the way we drive. They phrased their mission like this:

THE OLD PARADIGM

With high capital intensity, current global auto manufacturers design a single model, make hundreds of thousands of copies a year, and push it through a network of dealerships. Mass Customization and the search for low-volume runs is elusive and expensive. The customer feedback loop is inadequate and broken.

HOW WE DO IT DIFFERENTLY

We will license a lightweight, superior safety chassis that can be produced profitably at 2,000 units/year. On top of that we will layer design from our open-source design community. This community empowers an army of hotshot competitive designers from around the world to innovate and refine design. Our team specifies the target segment that fits the price point. The community delivers the innovation. These designs are then transferred to our network of suppliers who deliver the necessary subassemblies direct to the Local Motors facility on a just-in-time basis. All cars are assembled, tested for quality, and sold locally by a 20-person business unit at a facility with 1/100th the capital of today's auto plants.

One of the great advantages of building such a car today is that it plays into the global automotive manufacturing trends of the past three decades. All those shifts, led by the Japanese, from monolithic factories to an ecosystem of suppliers providing parts on a just-in-time basis, means that practically anything you need is on the market and easy to get. Small companies may not get the parts quite as quickly or as cheaply as Ford, but the global automotive supply chain is essentially open to all. It can work in units of millions and units of ones: yet another scale-free network, just like the Internet.

The thirty-eight-year-old Rogers favors military-style flight suits, an echo of his time as a captain in the Marines, including action in Iraq, and he boasts both a Harvard MBA and a stint as an entrepreneur in China. While at Harvard, Rogers saw a presentation on Threadless, the open-design T-shirt company, which showed him the power of crowdsourcing.

Cars are more complicated than T-shirts, but both are examples of "platforms" on which many people can display their talents and collectively innovate. And in both cases there are far more people who can design them than are currently paid to do so. In the automotive world, the majority of students who study car design don't get jobs in the industry; instead they end up designing toothpaste tubes or kids' toys. That makes them *frustrated would-be car designers*, exactly the pool of talent that might respond to a well-organized vehicle design competition and community.

A competition for every hubcap

Local Motors started in Wareham, Massachusetts, about an hour south of Boston, in an industrial park behind Factory Five Racing, a kit-car company and investor in the new firm. The kit-car connection is both a part of Local Motors' heritage and a warning of what it must avoid. Kit cars have been around for decades, standing as a proof-of-concept for how small manufacturing can work in the car industry. They combine hand-welded steel tube chassis and fiberglass bodies with stock engines and accessories. Amateurs typically assemble the cars at their homes.

In the kit-car business the vehicles are typically modeled after famous racing and sports cars, making lawsuits and license fees a constant burden. This makes it hard to profit and limits the industry's growth. Factory Five has sold only about eight thousand kits since it started in 1995.

Rogers and his cofounder saw a way around this. Their company would build only original designs; rather than invoking classic cars, they would reimagine what a car could be. The product would be created by its community, who are also its customers. But don't confuse a community with a committee. The winning designs would be decided by voting and competition, not compromise and consensus.

In 2008, Local Motors started its contest for its first car, a Baja racer. To help steer the community and seed their work, Rogers challenged them to use the World War II–era P-51 Mustang fighter plane as inspiration: it's a classic and gorgeous aircraft that represents some of the qualities he hoped the car would eventually display: power, toughness, agility, and cool. Most important, it wasn't already a car, so presumably the company wouldn't get sued for infringing someone's intellectual property with the homage.

The winner of the overall design was Sangho Kim, a graphics design student at the Art Center College of Design in Pasadena, California (he eventually claimed $20,000 in prize money for his contributions). But once his body had been selected, there were more than a dozen other competitions for subassemblies ranging from the rearview mirror to the stylish vinyl "skins" that substitute for paint on the body. What all the contributors had in common was a refusal to design just another car, compromised by mass-market needs and convention. They wanted to make something original—a fantasy car come to life.

In the end, more than 160 people contributed to the eventual design.

How to avoid the usual perils of committee design—either a camel or a gold-plated elephant? The Local Motors team exercises good old-fashioned leadership. At one point in the Rally Fighter design, the community fell in love with a taillight design of their own creation. Okay, responded Rogers, we can do that. But it will add $1,000 to the price of the car. Replied the community, "We don't love it that much!" They settled on a seventy-five-dollar part from Honda, which actually looks absolutely fine on the car. Rogers gently led the community into collectively getting smarter about car economics, without having to dictate the outcome.

It's worth pausing a moment and looking more closely at the community members, which now number some twenty thousand. They are a mix of amateurs and professionals, some already car designers, others designers from other fields, and yet others just car enthusiasts. They pick the problem areas they want to focus on, depending on what they know and what needs to be done: industrial design, dynamics, "skins," electromechanical systems, operations and sourcing, and others.

What they don't do is pull rank based on credentials. Amateurs have as much influence as professionals. The same is true in almost any open-innovation community: when you let anyone contribute and ideas are judged on their merits rather than on the résumé of the contributor, you invariably find that some of the best contributors are those who don't actually do it in their day job.

Rogers describes the participants as falling into two classes: "solution seekers" and "solvers." The first want something in particular done, and the second like to solve problems of any sort. Because it's an open-source community, people creating things for their own needs tend to post them, both in the development to get help and advice, and after they're done. And because there's so much of this in-progress posting, there's always something to help with if you're so inclined. What makes the community work is homophily ("love of the same"), the tendency for people to associate and bond with others like themselves in a network.

What this taps is the Long Tail of talent; in many fields there are a lot more people with skills, ideas, and time to help than there are people who have professional degrees and are otherwise credentialed. Exposing this latent potential, both of professionals looking to follow their passions rather than their bosses' priorities and of amateurs with something to offer, is the real power of open innovation.

Take the graduates of the Arts Center College of Design, which is one of America's top car design schools. It has about 180 students in its undergraduate transportation program, which is mostly about automobiles, and many hundreds of others in related fields such as industrial design. An estimated fifty of them will eventually work for car companies. Most of the others will get jobs designing some other kinds of products and working for consumer packaged-goods companies.

So most of the car design students won't design cars in their day jobs. Yet for many of them the dream of designing cars is still there. There just aren't enough full-time design jobs in the car industry for them. They have to do something else for a living.

But what the Local Motors community offers is a way to design cars even if it isn't your job. Those Arts Center students who don't end up in the car industry still have the necessary skills, experience, and ideas—they just aren't going to be paid to do that by day. But at night they can still design cars, following their heart. And if their design wins, they could even make some money, as Sangho Kim did.

What makes these new models so powerful is that they tap the "dark energy" (or, as writer Clay Shirky calls it, "cognitive surplus") that's been all around us already. It's the ultimate market solution: open-innovation communities connect latent supply (talent not already employed in that field) with latent demand (products not already economical to create the usual way).

And if you can prove that you're a great car designer in such a community, it might help you get a job actually doing that. Thanks in part to his Rally Fighter success, Sangho Kim did just that and now works for GM in Korea.

Once the Local Motors community settles on a design, the company's engineers make it manufacturable. They construct a jig on which to weld the frame tubes and carve molds for the fiberglass body parts.

Most other components are simply ordered from car parts suppliers such as Penske Automotive Group; the engines and transmissions can be bought straight from big car makers such as BMW and GM, who will sell to third parties. The axle of the Rally Fighter comes from a Ford F-150 truck; the fuel cap comes from a Mitsubishi Eclipse. This combination—have the pros handle the elements that are critical to performance, safety, and manufacturability while the community designs the parts that give the car its shape and style—allows crowdsourcing to work even for a product whose use has life-and-death implications.

The final assembly is done by the customers themselves under an expert mechanic's tutelage, as part of a "build experience" at the Chandler factory. At any given time, a half dozen Rally Fighters are being built in two rows facing each other. Each has a custom tool cabinet and a rack of parts next to it; the mechanic coach is always working with one team or another.

As a buyer, you spend two long weekends (six days in total) assembling the car. You don't need to have even so much as opened a hood before—you'll have learned well enough how to do it by the time you're done. The first lesson is how to properly attach a nut. First you use a torque wrench to tighten it precisely. Then you go too far and strip the bolt, so you can learn the difference between tight and over-tight. And so on for all the other fasteners and assembly techniques, for a foreshortened mechanic boot camp.

It's mostly assembly, rather than real manufacturing. The steel tubing frames are already made, having been welded in a back room by two workers earlier. So are the fiberglass body parts. The engine is an off-the-shelf 6.2-liter V8 made by BMW or GM, and the automatic transmission is similarly stocked. So, too, for everything else, from the dashboard gauges to the suspension. If you look closely, you'll recognize that the rearview mirror is the same one found on a Dodge Challenger and the steering wheel comes from a Ford F-150.[31]

Typically, a team will be two people—often a father and son—but if you want to build the car yourself, with the coach's help, you can.

All you need to do is put the parts together. When you're done, you can drive it home. Although the car is at its best racing through the desert and flying over bumps and ruts, it's fifty-state legal, thanks to the stock engines that have already been tested and approved by the Environmental Protection Agency. So you can drive it to the mall, too, if you don't mind the stares.

Because the customers make at least 50 percent of the car themselves, all sorts of regulatory hurdles fall away, much as they do with "experimental" home-built aircraft, which are exempt from most Federal Aviation Administration regulations on the grounds that the owners are well enough informed to protect themselves, or to at least understand the risks. The Local Motors cars don't have to be crash-tested and they don't have to be fitted with airbags. Uncomfortable with that? Then this isn't the car for you. But there are others who are just fine with that.

Liability and consumer protection rules are also relaxed when customers make their own goods. When something goes wrong with your Rally Fighter, you don't take it back to the "dealer" or wait for a recall. You built it, so you can fix it. After you finish the car and drive it home from the factory, you even get a toolbox with all the gear you need to repair the vehicle. You're also part of a community that's engaged and eager to help one another.

To walk around the factory is to see something that reflects both the past and the future of the car industry. It's the distant past, in the sense that these cars are built by humans, with wrenches and screwdrivers, much as the first horseless carriages were. There's not a robot in sight (aside from the CNC machines that cut metal in the back room), and there are no assembly lines.

But it's also the future: the open-source community approach means that designs are not just faster, cheaper, and better, but also come already market-researched (at least by the most avid would-be users). Products developed by a community are more likely to be embraced by one. Several more designs are in the pipeline, and the company says it can take a new vehicle from sketch to market in eighteen months, about the time it takes Detroit to change the specs on some door trim.

Local Motors proved this in early 2011, when the Pentagon's DARPA research agency ran a competition for an "Experimental Crowd-derived Combat Support Vehicle" (XC2V). Local Motors' community snapped into action and came up with a design within weeks, which was refined by the company's engineers. Three and a half months later the design had won, and a month after that Rogers presented it to President Obama. Of course, the competition was designed to favor Local Motors–style communities, but it's hard to believe that a traditional defense contractor could have even got the paperwork done in three and a half months, much less designed and built a new high-performance armored car from scratch.

Not your father's DIY

How revolutionary is this? After all, DIY cars have been around for decades, and the humble dune buggy kit, with a fiberglass body on a VW bug frame modeled after the Meyers Manx design, was a fixture in the 1960s and 1970s. An estimated quarter million dune buggy kits have been sold,[32] and they, too, use off-the-shelf car components and custom composite bodies, just like the Rally Fighter. They didn't change the world, they certainly didn't threaten the big car companies, and they never really took off.

So what's different now?

Nobody expects Local Motors to get huge or sell millions of cars; indeed, they've set a cap of only two thousand of each model (and they're nowhere near that on their first). There have always been niche car companies selling exotic machines to enthusiasts; this is, in a sense, just another one. Rogers describes it as filling in the gaps in the marketplace for unique designs. He uses the analogy of a jar of marbles, each of which represents a vehicle from a major automaker. In between the marbles is empty space, space that can be filled with grains of sand—and those grains are Local Motors cars.

And at nearly $75,000 per car, it's not cheap. And although the Rally Fighter is a high-performance racer, there are no great technological innovations, nor is it doing anything other cars haven't already done.

But Local Motors has created more than a car. It's also created an innovation platform, in the same way that Apple's iPhone is a platform for independent software developers to build a business around their own apps that run on it. Not only can Local Motors' community produce new designs faster, cheaper, and better than the conventional way of small teams working behind closed doors, but also, because the designs are all online and open, community members can create their own projects and businesses around them. So if you think it would be cool to add an automatic tire inflation system to the design, just do it. If people like it, have it made and sell it yourself. No need to go through Local Motors and lobby the engineers to add it for you; the car is an open design, co-owned by its community.

Indeed, in late 2011, Local Motors launched Local Forge as a specialized community to do just that.

"We'll continue to do the 'halo' projects," Rogers says, "but this platform is for everything else." Microfactories in San Francisco and Dallas are coming next to help build the community's designs.

Still, that's not so different from the third-party add-on markets that grew up around the old dune buggy designs. But what happens as cars become more like computers on wheels, driven by electric power systems and controlled by software? Then the notion of a "platform" becomes far more interesting.

The next market for Local Motors will be applying its model to electric cars. An electric car replaces the gas engine with electric motors on the wheels, replaces the gas tank with a stack of lithium-polymer batteries, and replaces all the mechanical aspects of a drive train with software. Anyone can buy motors and batteries, and, as the open-source phenomenon has proven, communities can often write software better than companies can. Now consider that electric cars are not stand-alone vehicles but are part of entire networks—the smart electric grid at home, the network of distributed chargers on the street, and the mobile phone networks, which they use to find chargers.

Whom do you trust to create great networked software and devices? You'll probably include Apple and Google on that list, along with any number of tech startups. But Toyota, Honda, Nissan, or even BMW and Mercedes probably don't come to mind.

The shift from cars-as-manufactured-machines to cars-as-rolling-computers is where the difference between yesterday's DIY cars and tomorrow's will really become clear. Sure, the Rally Fighter is not so different from the dune buggy. But the first Local Motors electric car will be something else entirely. And then the power of the community development model may be something the big car companies not only notice but envy.

GM's Volt took six years and $6.5 billion to develop. Tesla is an electric car company built on Silicon Valley entrepreneur lines, but its Roadster took six years and cost $250 million. Meanwhile, the Rally Fighter took eighteen months and cost $3 million. Granted, the Rally Fighter is much less complicated than the two electric cars I've compared it with. But as we enter the electric age, the complexity becomes mostly in the bits, not in the atoms. And there's no reason a smart community can't do that faster, better, and cheaper than any single company.

How might that change things? Well, for starters it can create an alternative to the notion of planned obsolescence and disposability. As products like cars become more about their software than their hardware, it becomes possible to reverse the arrow of time—they can get better after you buy them, not worse, because they can get software updates.

Think of how a website improves as the site's developers add new features and improve its design. Now imagine if your car did the same thing. The more it's powered by software, the easier that is.

Cars are, after all, increasingly driven "by wire," not by mechanical linkages (if you have a new car, odds are neither your pedals nor your steering wheel is physically connecting to the engine or wheels; they're essentially just joysticks that instruct software to actually move the vehicle). So why doesn't the car company constantly update that software to improve the car's performance, the way your computer Web browser is regularly updated?

The cynical answer is that the car company would rather you bought a new car. But a community-created product places no such premium on planned obsolescence. If people want to give older products new life, they can and do. New bits can bring new life to old atoms.

Ford, for one, is already paying attention. In early 2012, it worked with TechShop to bring one of the shared Making facilities to its home city. The Detroit TechShop is huge, at 17,000 square feet, and is stocked with $750,000 worth of laser cutters, 3-D printers, and CNC machine tools. Ford employees are free to use the space day or night for projects related to their work or personal projects, and Ford intends to give out 2,000 memberships in the first year. Ideas Ford employees have prototyped at the new makerspace, including a method for rocking a car out of snow, a one-way valve to let air out of a car to help with defogging, and a "kick plate" to help get in and out of test vehicles. Since the program began, patent submissions at the company have risen 30 percent, something for which its managers credit the TechShop injection of Maker spirit.

This is how industries are reinvented.

Detroit West (again)

You don't have to imagine what an entire car industry built along these lines might look like—it's already here. At the former GM/ Toyota NUMMI (New United Motor Manufacturing, Inc.) factory in Fremont, California, Tesla has built the most modern factory in the world. It happens to build cars, but it could build anything. It is not just automated, it's a veritable robot army. Hundreds of general-purpose KUKA robot arms do everything from metal-bending to assembly. Flat-topped robot vehicles carry car chassis around, charging themselves on inductive pads as they go. Robot painting arms from Fanuc can open car doors to spray around them, and then close them again when they're done.

Tesla will be making twenty thousand cars per year at this factory, which may sound like a lot, but still makes it a niche player in the global automotive business. But what's smallish for cars is still massive for everyone else. The Tesla factory occupies part of a building nearly a mile long. It will employ more than one thousand people. It is already the biggest factory in Silicon Valley. If you've seen the movie *Iron Man*, you'll have a feel for it. The film's protagonist, Tony Stark, was modeled after Tesla founder Elon Musk, and the factory looks like nothing more than the movie brought to life.

Part of what makes this factory so innovative is that these are not your regular cars. For a start, the Model S, which the factory will start with, is pure electric, which means that it shares as much with a laptop computer as it does with a traditional gas-powered car. Rather than complicated mechanical components such as an engine, transmission, and drive train, the Tesla cars have lithium-ion battery packs, electric motors, and sophisticated electronics and software. That means that they have a tiny fraction of the number of mechanical parts of a traditional car. They're simpler, and thus easier to build.

On a tour of the factory on its opening night, Gilbert Passin, Tesla's vice president for manufacturing, explained that the plant is like a massive CNC machine—it can be configured to make almost anything. The entire factory is programmable and every car can be different. The same plant can make several different models of cars simultaneously with totally different parts, even alternating among them. Henry Ford pushed standardization and "any color as long as it's black," but Tesla pushes customization, from the colors of the trim to the number of battery cells in the lithium pack. Tesla can even

road-test cars indoors on a special rumble track of various bumpy surfaces to detect loose or squeaky fittings. The track is next to the final
assembly line. If there are any problems, the people to fix them are
right there, which would be impossible with the emissions of internal
combustion vehicles.

The Tesla factory operates on a principle of manufacturing "units
of one," closer to the dream of mass customization than any automotive manufacturer has ever come. Because so much of the car is
made in the factory itself, there is no need for a big inventory of components or long supply chains and the inflexibility that comes with
them. With vertical integration comes total control—it's the ultimate
just-in-time process. It fabricates what it needs, when it needs it.

Contrast this with the GM/Toyota factory that previously oc-
cupied this space. In 1984, NUMMI was launched as an ambitious
effort to bring to American car-making the previous revolution in
production efficiency, the Japanese "lean manufacturing" techniques
that had been pioneered by Toyota. NUMMI itself was occupying
a factory that had failed: GM's Fremont Assembly site, which had
closed two years earlier after twenty years of operation as what was
generally considered the worst car factory in America. The GM plant
embodied everything that had gone wrong in the U.S. manufacturing
mode in the 1970s and 1980s, from outmoded technology to labor
unrest. It had it all: union corruption, a workforce that ranged from
apathetic to antagonistic, even drug dealing and prostitution in the
parking lot.

NUMMI was meant to help reinvent the American car industry,
starting from the factory floor. It was, in some sense, the first automotive "brownfield" site. Take a failed factory from the old era, replace
as much as you can, and begin again with a totally new way of doing
things—a "greenfield" strategy built on the grounds of an existing
plant. Japanese lean manufacturing was mostly about ways to bring
workers more into the process, encouraging them to give constant
feedback to eliminate waste and reduce errors. The hope was Ameri-

can factory workers could be made as productive as Japanese ones if given a better working environment that allowed them to take ownership of their output and tap their ideas on how to improve processes.

The parallels between then and now are striking. The ambition was the same: flexible, efficient, high-quality manufacturing, using automation to improve quality and just-in-time supply to lower costs and increase flexibility. But the difference is that then, automation meant custom-made automated handlers, each specialized for a single task, since powerful general-purpose robotic arms had not yet been developed.

The first generation of computer-controlled automation was closer to the steam loom than to a robot—it did one thing better than a human, but only one thing. As a result, it was efficient to make one product but incredibly hard to change the production process to make another. Before GM and Toyota closed NUMMI in 2009, it was making Toyota Corollas and Tacomas in different parts of the plant. There was a last-ditch proposal to save it by making GM-branded Prius hybrids there instead, but it was just too hard to change the plant.

Likewise, the NUMMI just-in-time supply model was far better than the batch-ordering of the traditional Detroit way, but it was still dependent on a long and complicated chain of suppliers, most of which were not based in California. Indeed, if anything killed NUMMI in the end, it was that the economics of having a plant so far from suppliers in the Midwest made less and less economic sense in an increasingly competitive market. Just-in-time made supply chains better, but they were still supply chains. The more dependent a factory was on parts made elsewhere, the less flexible it could be and the more it was exposed to the risk of disruption and pricing uncertainty. Because NUMMI was so dependent on an extended chain of suppliers, much of the factory was devoted to inventory and storing pre-made parts.

Today, the big difference is digital manufacturing. Unlike

NUMMI's custom automation, most of the Tesla robots are standard KUKA machines with light composite arms, six axes of movement, and the ability to lift 1,000 kilograms. Not only can they be reprogrammed for different tasks in just minutes, but they typically do dozens of different tasks as part of their regular job. Next to the KUKA arms in the assembly wing of the Tesla plant is a rack of different heads. An arm may start with an aluminum welding head, then switch that out for a bolt-driving head, then switch that out for a gripper, all automatically. Even the robots that simply move sheet metal from one stamper bay to another are KUKA arms. Unlike the custom transport machines they replaced, they use suction cups or other air-pressure graspers to carry material of any size and shape. Tesla's stamping machines were inherited from NUMMI (adapted to stamp light aluminum rather than the old steel), but the automation that drives them is all new.

So, too, for the supply chain. Musk is a zealot about bringing as much fabrication as possible in-house, and he's got the experience to know how to do it. This is what he did with his rocket company, SpaceX, which is now leading the private space industry. Its basic rocket technology is not much different from what NASA uses, but its production processes are what allows it to get to orbit at a fraction of the cost. Unlike the complex (and politicized) network of contractors, subcontractors, and sub-subcontractors of NASA's aerospace industry model, SpaceX makes almost everything itself using digital fabrication tools. Technology allows it to vastly simplify the complexity and bureaucracy of manufacturing, cutting costs by as much as a factor of ten and improving reliability. It doesn't need to reinvent the physics of space flight to improve on the NASA model; most of the innovation happens on the factory floor.

Tesla aims to do the same thing to the car industry. The old supply chains were based on the classic economic principles of division of labor and comparative advantage. The company that had the skills and tooling to make transmissions was not the same as the one that could make plastic dashboards or ABS braking software. Each specialized, and the buyers combined them all with supply chains.

This was like the early days of computing. There were specialized computers for accounting, others for ballistic missile trajectories, and yet others for the census. Then researchers invented the general-purpose computer, and today the PC on your desk can do anything. Each program you run reconfigures the machine for a different function. What your mouse does in a Web browser is different from what it does in the Call of Duty video game. Your computer can be a book, a phone, a television, a newspaper, a plaything, or a security guard, depending on what software it is running.

Likewise for the robotic factory. General-purpose robots can be reconfigured by software as easily as a PC. By using other general-purpose digital fabrication tools, from powerful laser cutters that create the stamp forms to shape metal to CNC machines that make the molds for plastic, Tesla can do much of what used to be outsourced to suppliers. By focusing on a product that is itself an outgrowth of the computer industry—the electric car, which is more digital than mechanical—the very parts that Tesla makes are reconfigurable. Rather than using a complex mechanical drivetrain, the performance of the Model S comes from software. Rather than a dashboard full of dedicated dials, most of the Tesla displays are on a single multipurpose screen, just like a PC.

What kind of manufacturing future does this allow? One that can let America and other relatively high-cost countries compete. Cheaper foreign competition and outmoded and inflexible labor-intensive production processes closed NUMMI. Now robotics are reopening it as Tesla.

The robots didn't replace humans in this case. NUMMI was gone and the factory was empty—there were zero jobs here, and everyone had lost. Instead, robots brought life back to a dead plant, and are bringing one thousand new jobs with them. These new jobs are higher-skilled and will pay better than the old ones. Yes, that means many of the workers from the old NUMMI plant will not have the skills to work at the new one, but some will. More to the point, this is a model that can stand up to the economic pressures of globalization and succeed.

Western companies can buy KUKA robots as cheaply as Chinese companies can. The labor component of products such as cars is falling rapidly as automation takes over, making the usual labor arbitrage economics less relevant. The raw materials—plastics, bauxite (aluminum ore), even lithium—are sold on the global market, and everyone pays more or less the same price. What's left is the cost of land, electricity, and taxes. Those are still more expensive in the West, but the gap is far narrower than what it was with labor. With the rise of the robotic factory, the multicentury global trade flows toward cheaper workers may be coming to an end.

To be sure, the Tesla factory is a special case. It got what amounts to a huge subsidy in its portion of the old NUMMI plant, which it was able to buy for just $43 million, complete with lots of functioning equipment. As a relatively new car company (it was founded in 2003), it didn't have to inherit the pension obligations and labor unions of the Detroit giants, nor did it face pressure to preserve jobs rather than automate. There's the small matter of the half-billion-dollar federal loan it got in 2010. And let's face it: it could still fail. It's trying to break into the car industry with an expensive vehicle using bleeding-edge pure electric technology in a world where even the giants are having trouble getting people to pay extra for decade-old hybrid technology.

But whatever happens to Tesla, its production model will triumph. It simply reflects the direction all advanced manufacturing is going, driven by the power of digital fabrication technology. It's no coincidence that the KUKA robots are made in Germany. Such flexible automation is why manufacturing in Germany, a high-cost country, has been able to thrive in the face of Chinese competition, making it the engine of the European economy. Tesla's factory is simply the newest to be built on this model, and thus the most innovative. Today it builds cars. But the same model could build anything.

Every few generations, the fundamental means of production is transformed: steam, electricity, standardization, the assembly line, lean manufacturing, and now robotics. Sometimes this comes from management techniques, but the really powerful changes come from new tools. And there is no tool more powerful than the computer itself. Rather than just driving the modern factory, the computer is becoming the model for it. Infinitely flexible and adaptable, general-purpose industrial robots can be combined to create the universal Making Machine. And like computers, they work at any scale, from the mile-long NUMMI plant to your desktop. That—not just the rise of advanced technology, but also its democratization—is the real revolution.

Chapter 9

The Open Organization

To make things a new way, you need to
make companies a new way, too.

In the mid-1930s, Ronald Coase, then a recent London School of Economics graduate, was musing over what to many people might have seemed a silly question: Why do companies exist? Why do we pledge our allegiance to an institution and gather in the same building to get things done? His eventual answer, which he published in his landmark 1937 article "The Nature of the Firm,"[33] was this: companies exist to minimize "transaction costs"—time, hassle, confusion, mistakes.

When people share a purpose and have established roles, responsibilities, and modes of communication, it's easy to make things happen. You simply turn to the person in the next cubicle and ask that individual to do his or her job.

But in a passing comment in a 1990 interview, Bill Joy, one of the cofounders of Sun Microsystems, revealed a flaw in Coase's model. "No matter who you are, most of the smartest people work for someone else," he observed, stating what has now come to be known as "Joy's Law." His implication: for the sake of minimizing transaction costs, we don't work with the best people. Instead, we work with whomever our company was able to hire. Even for the best companies, that's a woefully inefficient process.

In a sense, Joy's quip was simply a modern reflection of the work of a Coase contemporary, Friedrich Hayek. While Coase was explaining why centralized organizations exist, Hayek was arguing that they shouldn't. In his own landmark paper in 1945, "The Use of Information in Society,"[34] Hayek observed that knowledge is unevenly distributed among people and that centralized planned and coordinated organizations would be unable to tap distributed knowledge (his point: only free markets could).

A half century later, when Joy made his similar observation, Sun Microsystems was one of the hottest tech companies in the world. His remark was a warning not to become complacent about that. Even though Sun thought it had the best engineers and the best technologies, there were more good people outside the company than within. Regardless of what Sun did, the competition from outside the company would always have the potential to be greater; open innovation would beat even the strongest individual companies. And indeed, Sun was eventually eclipsed and is no longer an independent company (it's now a division of Oracle, and Joy left to become a venture capitalist).

The same is true today. Take even the best company you can think of, say Apple, and consider how it hires. First, it's based in the United States, and most of its employees are in Cupertino, California. So there's a bias toward those who are already in the United States, or can legally work in the country, as well as toward those who live in the San Francisco Bay Area or are willing to move there. (It's lovely in Cupertino, but if your spouse doesn't want to leave her family in Rome or Chang Mai, that may matter more.)

Like all companies, Apple favors people with experience in the industry it's hiring for, and it likes to see degrees from good universities as an indication of intelligence and work ethic. Even though Steve Jobs was a genius teenage dropout, there aren't many others like him at Apple. The company may "think different," but these days it hires pretty much like every other good company: based on professional qualifications.

It also can only hire people who want to be hired. So that eliminates all those elsewhere who love the jobs they're already in and don't want to leave. It tends not to hire children, the elderly, and felons, regardless of how smart they may be. Also, anybody who can't keep a secret and doesn't want to be bound by the terms of an employment contract, and so on.

Yet there are smart, even brilliant, people who fall into all those excluded categories. By being a company, rather than an open-ended community, even Apple is subject to Joy's Law.

Communities tend to be more egalitarian, in part because they typically don't have the same legal obligations and risk as a company. They don't have to check references and get people to sign contracts before they participate, the way a company typically must. They can afford to take more chances with participants, because the consequences of things not working out are so much smaller when you're not promising people a wage (not to say they can't get paid for work done, but any rewards tend to come after the fact, not as a salary).

To be sure, communities can't do everything and the world's economy can't run entirely on volunteerism. But Joy's point was that labor markets are changing. With the Internet, you don't have to settle for whoever is sitting in the next cubicle. You find and tap the best people out there, even if you're in Detroit and they're in Dakar. Or, more to the point, they can find you. In open-innovation communities, participants self-select. They are drawn to cool projects and smart people, and when work is done in the open, they can find it. I learned this firsthand in my own robotics community.

A most unlikely CEO

A few months after I'd launched DIY Drones and had a few hundred members, a guy named Jordi Muñoz signed up and posted a link to a cool hack he'd done with a new open-source microprocessor board called Arduino: he'd figured out how to use it to fly a toy helicopter with a Nintendo game controller.

His first forum post began this way: "English is not my first language, sorry if I made mistakes trying to describe this project. I made an autopilot for my RC helicopter with accelerometers extracted from the NunChuck of Nintendo Wii." He included some pictures of the helicopter, now augmented with circuit boards and a tangle of wires, and, shortly thereafter, a video of it actually in the air.

People quickly took notice. Another poster responded with encouragement: "Your English is very good; don't worry too much about translations; a picture is worth a thousand words, and we're excited to see [the] video. That's an excellent helicopter you put together. It's cool that people are coming up with complex ideas and getting them to work."

I was impressed, too; I'd never used Arduino, but this prompted me to look more closely at it. I got in touch with Muñoz to ask some more questions about Arduino, and we started a friendly correspondence. I liked his energy and was impressed by his fearless experimentation and effortless grasp of software concepts that I had struggled to understand. I had a feeling that he was on to something; his instincts kept leading him to more and more exciting technologies, from sensors he found and figured out how to use to algorithms he tracked down in obscure papers.

Eventually, we started to do some projects together on DIY Drones—first an airplane autopilot and then an autonomous blimp controller board. We'd trade circuit-board designs back and forth and we both spent our evenings hunched over soldering irons on our respective worktables, attaching components and testing them. He taught me how to program Arduino and the best places to buy components and get boards made. I wrote the blog posts describing our progress and documented the projects with online tutorials.

Initially, we were just electronics hobbyists sharing hacks with other DIYers. We'd upload links to the places to get the parts to follow along with our projects, but if you wanted to do it you'd have to have your own printed circuit boards manufactured for you and you'd have to buy all the components from online suppliers yourself. As a result, only a few dozen other community members were using our designs.

It was clear that if we wanted more people to participate in these sorts of projects, we'd have to make it easier for them. Rather than share design files, leaving members to their own devices to actually buy the parts, we should offer kits with everything included. And that meant buying gross lots of the parts, packaging them up in kits, and finding some way to take orders.

That, in turn, meant starting a proper company. I asked Jordi to join me as cofounder. And when he agreed, I thought that might be a good time to ask him a bit about himself.

Here's what I learned: At the time of his first posting, Jordi Muñoz Bardales (his full name) was nineteen years old. He was a native of Encenada, Mexico, and had gone to high school in Tijuana. He had just moved to Riverside, a suburb of Los Angeles. His high school girlfriend, who has dual citizenship, was pregnant, so they had recently got married. He was playing with the helicopter in their Riverside apartment because he didn't have anything else to do while they waited for his green card. He had never been to college.

Needless to say, none of that mattered. The only thing that mattered was what he could do, which he had already resoundingly proved. Today, Jordi is CEO of 3D Robotics Inc., a multimillion-dollar company with a state-of-the-art factory in San Diego. As I write this, he is twenty-four years old.

How did this transformation happen? Three steps:

1. A smart kid who didn't happen to be born in the United States, didn't speak great English, and didn't do terribly well in school *did* have access to the Internet. Because he was curious and driven, he used the greatest information resource in history to make himself one of the world's leading aerial robotics experts. He was just following his passions, but in the process he got what amounts to a "Google Ph.D."

2. When I decided, against all odds, to start a company to do aerial robotics, I did it with the smartest guy on the subject I knew. I didn't ask for a résumé. It wasn't necessary. That guy had already proven himself by making extraordinary things.

3. With a lot of support from the community, some fearlessness, and once again the power of Google research, Jordi learned the fundamentals of electronics manufacturing and manufacturing operations. He hired a smart team of mostly other twentysomethings, a mix of Americans and bicultural Mexican engineers from Tijuana.[35] They did the same thing, quickly learning everything they needed to know online, both in research and by asking people. Eighteen months later, they were running a world-class robotics factory.

Twenty years ago, what would have been the chances that when the editor of *Wired* magazine decided to start an aerial robotics company, he would end up partnering with a nineteen-year-old high school graduate from Tijuana? Yet today it seemed like the most natural thing. Why wouldn't you start a company with people with whom you were already working well, who had already proven their mettle? It seems so much riskier to take a flier on someone you don't know, just because that person has a degree from a good school.

This is the Long Tail of talent. The Web allows people to show what they can do, regardless of their education and credentials. It allows groups to form and work together easily outside of a company context, whether this involves "jobs" or not. And these more informal organizations are much less constrained by geography; talented people can live anywhere and shouldn't have to move to contribute.

As *New York Times* columnist Thomas Friedman puts it, "It used to be that only cheap foreign manual labor was easily available; now cheap foreign genius is easily available." Not just cheap because they work for less money; cheap because they're often working for no money at all, as global volunteers in a project that they believe in while some other job puts food on the table.

Today our robotics company has about a hundred contributors whose work has made it into a product. About twenty of them are paid employees, mostly working on hardware engineering and manufacturing in the factory. The other eighty are volunteers working on software. The volunteers all have other jobs, ranging from an Apple engineer to a cake maker, but some of them put in what in some weeks amounts to full-time work on the robotics projects. Some of them are professional software programmers just looking for a new challenge; others are amateurs who have made this their hobby and taught themselves what they needed to know.

Perhaps, if this was a company crafted in the Coase model, we would have found and hired some of the first category—the professionals already active in the field. But we certainly would have missed the cake maker, the graphics artist working for the Brazilian ad agency, the guy who runs the Italian ambulance radio company, the retired car-dealership owner, the Spaniard working for an energy company in the Canary Islands, and all the others who followed their passions into the project, even though their careers had taken them elsewhere.

In short, because we don't operate the company in a Coaseian model, we've got more and smarter people working for us. *We minimize transaction costs with technology, not proximity.* A social network is our common roof. Skype is the "next cubicle." Our shared purpose is really shared, not dictated.

Joy wins: The open-manufacturing model

Joy's Law and the new breed of companies and communities built on open-access Web principles turned Coase's Law upside down. Now, working within a traditional monolithic company of the sort Coase had in mind often imposes *higher* transaction costs than running a project online. Why turn to the person who happens to be in the next office, who may or may not be the best person for the job, when it's just as easy to turn to an online community member from a global marketplace of talent?

Companies are full of bureaucracy, procedures, and approval processes, a structure designed to defend the integrity of the organization. Communities, on the other hand, form around shared interests and needs, and have no more process than they require. The community exists for the project, not to support the company in which the project resides.

Yet communities can't make physical goods by themselves. Somebody has to do the manufacturing, handle the inventory, get the liability insurance, and run the customer support, and that takes money, a legal structure, and real day-to-day responsibilities. Thus, a company.

So, in the new manufacturing model, you need a new kind of manufacturing company, too. At its core, it has to incorporate all the skills and learning of traditional manufacturing companies—tight quality control, efficient inventory management, and supply-chain management—so that it can compete with them on basic price and quality. But it also needs to incorporate many of the skills of Web companies in creating and harnessing a community around its products that allow it to design new goods faster, better, cheaper. In short, it must be like the best hardware companies *and* the best software companies. Atoms *and* bits.

Maryam Alavi, vice-dean of Emery University's Goizueta Business School, argues that the only way firms can continue to have lower transaction costs than the open market is if they become more complex *internally* in order to respond to the increasingly complex *external* market. In the Aspen Institute's "The Future of Work," she explained that this was due to the "law of requisite variety" in systems theory, and she argued that a system must be as complex as the environment it is working within: "There are parts of the organization that are going to become more hierarchical because of the uncertainties that they deal with or don't deal with. And there are parts of the organization that will need to be highly dynamic, open, and changing."[36]

Thus the new industrial organizational model. It's built around "small pieces, loosely joined." Companies are smaller, virtual, and informal. Most participants are not employees. They form and re-form on the fly, driven by ability and need rather than affiliation and obligation. It doesn't matter who the best people work for; if the project is interesting enough, the best people will find it.

The open supply chain

How would an American manufacturing economy built on such principles look?

On the face of it, that's a pointless question: read the daily headlines, and you'd hardly be blamed for thinking that there's no future for American manufacturing at all. After all, it's more than just that labor costs are lower elsewhere. Even more important, the ecosystem of suppliers and skills has moved abroad, too.

As Garry Pisano and Willy Shih point out in a telling 2009 *Harvard Business Review* article on American competitiveness,[37] Amazon can't make a Kindle 2 in the United States because:

1. The flex circuit connectors are made in China because the U.S. supplier base migrated to Asia.

2. The electrophoretic display is made in Taiwan because the expertise developed from producing flat-panel LCDs migrated to Asia with semiconductor manufacturing.

3. The highly polished injection-molded case is made in China because the U.S. supplier base eroded as the manufacture of toys, consumer electronics, and computers migrated to China.

4. The wireless card is made in South Korea because that country became a center for making mobile phone components and handsets.

5. The controller board is made in China because U.S. companies long ago transferred manufacture of printed circuit boards to Asia.

6. The lithium polymer battery is made in China because battery development and manufacturing migrated to China along with the development and manufacture of consumer electronics and notebook computers.

According to Pisano and Shih, only Apple "has been able to preserve a first-rate design capability in the States so far by remaining deeply involved in the selection of components, in industrial design, in software development, and in the articulation of the concept of its products and how they address users' needs." And even it still manufactures in China.

That's depressing. But let's remember that the American manufacturing industry, despite the gloom of the past few decades, is still the largest in the world (although it will soon be passed by China). U.S. factory output, in inflation-corrected dollars, has more than doubled since 1975 and is currently near its all-time high.

What's still being made in the United States? A combination of big things that will be sold in the country (such as cars), high-value items where the cost of labor is small compared to the price (such as airplanes), and specialty goods, where there is little commodity competition (such as medical equipment).

Companies such as General Electric, Procter & Gamble, 3M, Boeing, and Lockheed Martin, and even stalwarts such as US Steel, remain among the biggest manufacturers in the world. U.S. automakers, such as Ford and GM, are staging a remarkable turnaround (thanks in part to government intervention and tough reforms). Along with foreign companies that make cars in the United States, total output in 2011 was close to an all-time high, short only of the two years of the NASDAQ bubble of 2000.

So Factory U.S.A. is still working in some industries, despite the rise of China.

What that tells us is that there's more to the geography of manufacturing than simply a race to the cheapest labor rates. Being closer to the consumer means a company's design can fit their needs better, as Apple proved. Although an iPhone says DESIGNED IN CALIFORNIA. MADE IN CHINA on the back, 2011 research by Kenneth Kraemer of the University of California, Irvine, and two other American economists found that more than half of the price of the phone stays in the United States. They write:

> While these products, including most of their components, are manufactured in China, the primary benefits go to the U.S. economy as Apple continues to keep most of its product design, software development, product management, marketing and other high-wage functions in the U.S. China's role is much smaller than most casual observers would think. Add to that the increasing cost of transportation across the seas, the political risk of trade wars and tariffs and the hidden costs of delays and disruption in shipping along with the excess inventory needed to buffer that, and you can see why the Eastward migration of manufacturing may have peaked.[38]

Can Makers make jobs?

But one thing we have not been making more of in recent years is manufacturing *jobs*. Even as output doubled over the past four decades, manufacturing employment fell by about 30 percent over that period. The increased output was a result of improved production efficiency (mostly automation) leading to greater productivity per employee, not more workers.

Meanwhile, the biggest creators of jobs in the American economy are small and medium-sized business—exactly what manufacturing moved away from over the previous decades as companies searched for economies of scale to compete with low-cost labor overseas.

Actually, my statement above about small businesses creating jobs is not quite true. It's actually more correct to say that small businesses *destroy* a lot of the jobs that they create, since most small businesses fail before their third year. Even those that do survive are actually just sole proprietorships, which is to say a one-person show, and often not even full-time at that.

What really creates jobs is small businesses that grow into larger businesses. But unlike in the First Industrial Revolution, these don't have to be industrial giants with armies of workers. Most of the Internet economy is made of companies with a few hundred employees, like Twitter or Tumblr. The same is true for manufacturing companies that grew up along the Maker model.

Take, for example, Aliph, which makes Jawbone noise-canceling wireless headsets. Aliph was founded in 1999 by two Stanford graduates, Alex Asseily and Hosain Rahman, and now sells millions of headsets and portable JamBox speaker systems each year. It has no factories and outsources all of its production. Aliph makes bits and its partners make atoms, and together they can take on Sony.

Yet although more than a thousand people help to create Jawbone headsets, Aliph has just over one hundred employees. Everyone else works for its production partners. So, too, for most of the other successful companies that have followed this path. Although the revenues and profits outgrew the category of "small business," employment did not. Because these companies are built along Web lines, they tend to be lean.

But they also tend to be numerous, since the barriers to entry are so low. And with that many small manufacturers and companies, the odds that some of them will get big increase. The Silicon Valley model—that all startups are created with the hope of becoming the next Facebook—is what's really the engine of economic growth. Even though almost all of them will fail to reach those highs, if a few do they can create multibillion-dollar industries and tens of thousands of jobs.

And companies built on the Web-driven Maker model can do that. Why? For three reasons:

First, because most start with an open community, they have the powerful growth potential of network effects built in. The communities can not only provide a faster, better, cheaper product-development process, but they also often offer a better, less expensive form of marketing. Word of mouth is the best way to sell anything, and what's better word of mouth than the word of people who had a part in the creation of a product, or at least witnessed it?

Second, because these companies are built along Web lines, they're good at using the Web for everything, from finding low-cost suppliers to virtual manufacturing using service bureaus. Web-centric companies are simply better at using the best tools out there to save money and speed product development.

Finally, because they were born online, these companies are also born global. They typically serve a niche that cuts across national borders. As such, they are designed to be exporters from the start. They typically sell online, so they're not constrained by traditional distribution and geography. That means that they not only can grow faster, but can also fend off competition more easily—they're already competing on a global stage, so it's hard for imports to undermine them.

Meanwhile, the traditional threat of competition from low-cost-labor countries may not be as daunting as it once seemed. China, for starters, is getting more expensive. Wages in the industrial provinces such as Guangdong are rising at 17 percent per year, and the creeping revaluing of the yuan only makes that worse in real terms. American workers are also up to three times more productive (not because they're necessarily more skilled or harder-working, but because they tend to be matched with more automation, which amplifies individual productivity). The Boston Consulting Group estimates that the net cost of manufacturing in China will be the same as that in the United States by 2015.[39]

And as factory automation becomes more powerful, the labor component of the average product drops. And that means that the traditional labor arbitrage arguments for moving manufacturing jobs overseas will diminish. Right now, in the automotive industry, labor represents less than 15 percent of the cost of the vehicle (the United Auto Workers union claims that it is just 10 percent, but that includes only assembly-line workers, not office, management, and R&D). Robots are going to become only better and more numerous: a factory job increasingly looks like a decreasing number of workers making sure the robots get the components they need on time, and a shipping department.

The labor arbitrage view of global trade, a model that goes back to the dawn of the First Industrial Revolution, assumes that manufacturing will always flow to low-cost countries. But the new automation view suggests that the advantages of cheap labor are shrinking while other factors—closeness to the ultimate consumer, transportation costs (including possible carbon taxes), flexibility, quality, and reliability—are rising.

Caterpillar, for example, is tripling its excavator operations in Texas, adding another five hundred manufacturing jobs, because Texas is closer to its customers and supply chains. NCR is bringing its ATM production back from China to Columbus, Georgia, so it can get to market faster and improve internal collaboration. And even toymaker Wham-O is bringing back half of its Frisbee production from China, thanks to increasingly automated and efficient U.S. factories.

Meanwhile, niche manufacturing companies focus on being close to their customers, offering custom or quick-turnaround goods to customers who are willing to pay for that. One of the concepts that's taking off among regional development experts, whose job it is to attract businesses to their towns and cities, is the idea of "economic gardening." In the same way that small-plot gardens can thrive even in the presence of factory farms, small manufacturing companies can thrive if they are nimble and innovative.

In New York City, smaller companies still manufacture everything from envelopes (customers can easily visit the factory to inspect designs before they go on the line) to handcrafted BMX bikes at Brooklyn Machine Works (at as much as $2,800 a frame, cheap labor is not the priority). In San Francisco, a thriving group called SFMade represents scores of entrepreneurial manufacturers who trade on their locality, from Timbuk2 bags to Mission Motors electric motorcycles.

The sorts of businesses that capitalize on being close to their market range from custom furniture, which needs close contact with customers, to high-end mattresses (build-on-demand lowers cost), to niche couture (my own office building in the hot high-tech district South of Market also houses some textile factories, with immigrant Chinese workers working on locally designed clothes). That's always been the case, but now these companies aren't just local. If they're sufficiently innovative, they can sell globally, too, online.

Just consider the high-end chocolate made by San Francisco's Tcho, in a full beans-to-bars chocolate factory run on a converted pier on the Bay by the original founders of *Wired*. They started local, serving the same boutique demand for artisanal products that saw the rise of high-end coffee chains such as Peets (another San Francisco native) decades earlier. But because they're a product of the Web Age, they went global more quickly, both through e-commerce and online word of mouth. Now, five years after its founding, Tcho is sold by more than four hundred retailers around the country. The factory on the pier in San Francisco run by Web pioneers makes chocolate around the clock to keep up with demand.

The calculus of geography

I don't want to suggest that companies won't continue to outsource manufacturing to China or other low-cost countries. For many industries, the combination of relatively cheap labor and the concentration of suppliers that you can find in Guangdong is unbeatable. That is why no mobile phones are made in America, and why China is the toy capital of the world.

But what's clear is that it's not the only choice. At some scales, manufacturing in huge Chinese factories may continue to be an unbeatable answer. But at other scales, the advantages of making things close to home, with minimal delays and maximum flexibility, can be a better choice. And with more automation, the economic gap between manufacturing in China versus manufacturing in the United States is shrinking.

Here's a rough sense of how that "make it here" versus "make it there" calculus can look:

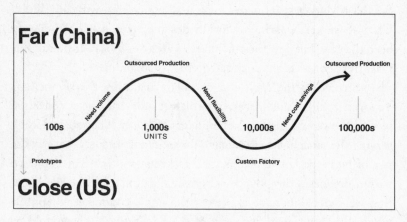

Imagine a new company, WindCo, making its first product, a small backyard wind turbine power generator. They make the first prototype themselves, as well as a handful of others to send to partners. Next, it's time to go into production. But because WindCo is small, they don't have that sort of manufacturing capacity themselves, so they outsource to a Chinese factory to do it for them.

That works to get the product into the market. But once sales take off into the hundreds, the limits of that model become clear. First, it's inflexible: when the product sells out, it takes months for a new supply to arrive, and also the Chinese factory prefers to work in large batches, so WindCo has to take delivery of huge shipments that it can sell only gradually, over time, leaving much of its cash locked up in inventory waiting to be sold.

Simple economics start to argue for local manufacturing. So WindCo sets up its own local factory, where it can make the turbines on demand. It is now much easier to manage the company's inventory and to make improvements in the product based on customer feedback and demand.

But let's say sales continue to rise into the tens of thousands. At that point China starts to look more attractive again as a manufacturer. The 30 percent price difference between making it locally and making it in Guangdong—which was not as critical as time and flexibility at a smaller scale—now is impossible to resist. This is even more true if a competitor enters the market at a lower price, and you have to compete on cost. Back production goes to China.

And so it goes. Companies can increasingly move manufacturing to wherever it makes most sense. They can do so because the design files are digital, the tooling costs of setting up a new manufacturing operation are minimal, and they all use the same robotic machinery, which can be bought anywhere.

This is a world where America *can* compete. And so can China. And Germany and Mexico and Poland. Digital manufacturing levels the global playing field. *Any* country can make things. The question is only what can they make better than anyone else.

A very modern factory

Look hard enough and you'll see examples of this everywhere. In Silicon Valley, of course, but also in places where you might not expect advanced manufacturing: converted car-repair shops in Brooklyn, industrial parks in suburbs of Las Vegas, farm towns in the middle of Wisconsin. What they all have in common is that they're located where the entrepreneurs who started these companies wanted to live. They didn't need to locate near railway tracks or highways, the way factories once did, nor did they need lots of land and cheap labor. Manufacturing can increasingly be done anywhere FedEx and UPS will pick up from.

Take Sparkfun. In 2003, Nathan Siedle was an undergraduate engineering student at the University of Colorado, in Boulder, an upscale college town an hour from Denver. He was finding it difficult to locate electronic components that he needed for his projects, but eventually he tracked down some suppliers online. At that point he might have declared victory and simply finished his studies. But like many of the Makers in this book, he decided to share his discoveries instead. He set up a small Web storefront to sell the hard-to-find parts, maxing out his credit cards on inventory in the process. He called it Sparkfun, a winking reference to the frequent experience of touching the wrong parts together and watching them fry in a shower of sparks. On the day he got his official state sales-tax exemption, making him a "real company," he was so excited he sped home on his motorcycle and got a speeding ticket serious enough that it required a court visit.

By the time Siedle graduated, Sparkfun had grown into a real company. Rather than getting a job elsewhere, Siedle decided to make a go of it. And since he liked Boulder, he just rented some space on the ground floor of a building in a local office park and set up shop.

Today, Sparkfun has more than 120 employees and annual revenues of around $30 million, and is growing by 50 percent a year. A basketball-court-sized ground floor is dominated by robotic electronic production lines, running day and night. Daily blog posts and tutorials have turned its retail website into a high-traffic community, with more than fifty thousand visitors a day.

Remember, this is in high-cost Boulder, Colorado, one of the most expensive real-estate markets in America. And the sector is electronics, a market that many had thought was lost for good to China. How does Sparkfun compete with such low-cost production? With automation, close ties to their customers and their needs (its hobbyist roots give Sparkfun great geek cred), and a community built around its website's daily tutorials, posts by its employees (now mini-celebrities in the Maker world). Sparkfun is proof that manufacturing success is not just about finding the cheapest labor.

When you ask people about the state of American manufacturing, they often quote the same depressing statistic: despite the U.S. clout in mobile phones, including Apple's iPhone, Google's Android phones, Motorola, and others, none of those phones are made in America. We may be technology leaders, but we can only make the bits (the product concept and software), not the atoms (the physical phones themselves). "Designed in California. Made in China" indeed.

But visit the Sparkfun factory and you'll get a different view. Unlike most of its bigger electronics parts supplier competitors, Sparkfun makes most of what it sells—right in Boulder. It has several large pick-and-place robot machines that place chips and other components in precise position on printed circuit boards faster than the eye can see. A conveyor belt takes the "populated" boards into an automatically controlled oven to melt solder paste under the chips, cementing them to the boards. Other computer-controlled machines load components and prepare boards. Three workers watch over the operation, which runs around the clock.

In short, electronics can be made in America, as long as they're specialty electronics, selling in the thousands, not millions.

The Kindle 2 and iPhone need the latest screens and fastest memory chips, which are made in volume by only a few manufacturers in Asia. But most other electronic devices don't need the very latest, smallest, lightest, and fastest parts. Think of electronics more like a smart thermostat in your house or the dashboard of your car. They don't need Apple-like performance. Instead, they get their value from the software that runs on their commodity parts. That's the sort of thing you can do anywhere.

Such specialty goods usually command higher margins and are less likely to face competition from other commodity suppliers. It's a classic market niche for a midsized manufacturing business. Big enough to sell globally and have an established brand, but not so big that it falls into the commodity deathtrap of razor-thin margins and scary overexposure to economic swings and the changing taste of fickle consumers.

By contrast, China's Foxconn, which makes Apple's iPhone and many of the other mass-market electronics you buy today, has about a million workers, which makes it the second largest non-state company in the world (after Wal-Mart) by employee count.[40] It runs entire company towns, and its workplace conditions (including suicides) make headline news. Foxconn doesn't develop its own product; it does outsourced manufacturing for others. But that means tiny margins. Economists estimate that it gets only $6.50 for the work of assembling a phone that sells for $300.[41] Likewise for most of the Asian suppliers that make the components that go into an iPhone. The lion's share of the profit goes to Apple, the designer. Which business would you rather be in?

Sparkfun, on the other hand, both designs and makes most of its products. And it does so in exactly the model I've described above: an open-innovation process built around a community of its customers. Most of Sparkfun's products are open-source hardware, which is to say that their design files are openly shared and can be modified. Many of them were actually designed by customers and simply reviewed and improved by Sparkfun engineers to make them easier to manufacture.

It's a classic community-centric company. The front of its website features not products but its blog, with chatty tutorials and videos from its employees. Its forums are full of customers helping one another. Every year Sparkfun throws an autonomous vehicle competition, featuring a live band playing robot-themed songs of its own composition, and lots of kids chasing self-driving cars (I've been competing in the aerial category every year since it started—no wins yet). At Maker festivals around the country, Sparkfun engineers teach people how to solder, which is actually a lot more fun than it may sound.

Sparkfun's employees are young, passionate, and appear to totally love their jobs. Dogs and hobbies are indulged at work (although not on the production floor); tattoos and indie punk rock reflect its culture. It's about as far from the "dark satanic mill" vision of manufacturing as you can imagine.

This is a twenty-first-century American Maker manufacturing success story. It thrives in the face of Asian competition. It's growing fast and creating jobs. It is very profitable. Equally important, it's got a great multiplier effect. Each conventional manufacturing job is typically credited with creating four other jobs in the community. But Sparkfun, because it sells technology that helps others build their own companies, has an even higher multiplier than that.

How high? It's hard to say, but here is one example: Facebook has about 2,500 employees as of this writing. But its chief operating officer, Sheryl Sandberg, estimates that more than thirty thousand people make their primary living as part of the "Facebook ecosystem," all the companies and services built on Facebook, from Zynga games like Farmville to all the "social media experts" hired to help companies navigate Facebook. That's at least a 10x multiplier.

Pisano and Shih, in their *Harvard Business Review* article on American competitiveness, called for a rebuilding of an "industrial commons"—the collective R&D, engineering, and manufacturing ability that can sustain innovation. Not just the ability to make stuff, but also the ability to invent it, the ability to make the parts that go into it, and the ability to train the generation who will do all that.

Successful technological companies can do this. Their trickle-down effects are not measured in dry cleaners and local pizza franchises serving their workers' families, but rather in the tools they sell that make other companies around them more powerful. In other words, they are not just creating new jobs, but creating *new companies* that create more jobs. Sparkfun, a very modern factory, is the hub of one such new industrial commons. The question is only how far this Maker Movement commons can spread.

Chapter 10

Financing the Maker Movement

**Where does making end and selling begin?
In the new Maker markets, it's often the same thing.**

Don't put a jellyfish in a regular fish tank. Just don't. What happens if you do is not pretty. First, it will be gently, inexorably, drawn by currents to the sides and corners of the tank, in particular the side with the intake to the pump. It will then be sucked into the pump itself, in which it will become wedged. And then the pump will rip it to pieces.

You may be tempted all the same. Jellyfish are perhaps the most beautiful, magical creatures you can have in a tank, as you will have seen if you've visited a jellyfish display in one of the larger public aquaria. Illuminated with colored lights, they're a moving art display, gently undulating in groups or peacefully alone, an ever-changing living lava lamp. But if you want one in your own home, you'll typically need a custom tank made at a cost of thousands of dollars.

This didn't seem right to Alex Andon. He had taken a fancy to jellyfish while sailing in the British Virgin Islands as a teenager. After graduating from Duke with a biology degree in 2006, he came to the San Francisco Bay Area for a biotech job. But the jellyfish fascinated him more, in part because San Francisco Bay is one of the best places in the world to catch them. He decided to quit his job and set up a company in a friend's garage to make custom jellyfish tanks. He called it Jellyfish Art, and it grew quickly, offering modified fish tanks with special pumps and custom water-flow systems that kept the jellyfish off the sides. He learned how to freeze plankton to make perfect jellyfish food, and how to ship small moon jellyfish live through the mail.

But as jellyfish pets grew increasingly popular, Andon decided he needed to design and manufacture a brand-new kind of tank, one designed from the ground up for jellyfish. It would have a laminar-flow filtration system, so there were no strong currents to trap the animals, and it would be lit with LEDs in colors that could be changed by remote control, for maximum visual effect. It would be small enough to place on a desktop, but large enough to hold four jellyfish without crowding.

This meant getting into the manufacturing business at scale, which would not be cheap. Normally, at that point, an entrepreneur would seek funding. A bank loan is one method; venture capital is another. But neither is easy to get, and both come with risks and loss of control. A bank loan would probably be collateralized with whatever property Andon had, and would have to be paid back with interest, while a venture capitalist would want a sizable portion of the company.

There was, however, another way. Over the past few years, a new phenomenon of "crowdfunding" has taken off, by which supporters and potential customers collectively contribute the money necessary to get the product made. Crowdfunding may take many forms, from glorified tip jars to formal loans backed by people, not banks.

The one Andon chose was Kickstarter, a website where people post descriptions of their projects and anyone can chip in to help. Rather than just making a donation, most contributors essentially preorder the product by making a contribution above a certain level. In the case of the Desktop Jellyfish Tank, donors who gave $350 or more would be the first to get the tank when it was available, at a lower price than regular customers would pay.

Kickstarter required Andon to set a minimum amount to be raised. If he hit that target within thirty days of posting the project, everyone who pledged money would have their credit cards charged, Andon would get the money, and he would be expected to go ahead with the project. If he didn't reach that amount, nobody would pay a penny and Andon would have to find some other source of funding. He set a target of $3,000.

The Desktop Jellyfish Tank hit that figure in less than twenty-four hours. And then it kept going. More and more donors poured in, thanks to word of mouth and pent-up jellyfish demand. By the time the thirty days expired, Andon had raised more than $130,000 and 330 people had preordered a tank. Andon was amazed and delighted; he had hoped that loads of people would want jellyfish in their homes, but he had no way of knowing for sure. Now he had the evidence: people were voting for his product with their wallets.

Andon now had seed capital to start production. He had guaranteed orders. And he had the confidence in knowing that the world wanted what he was making. And all this without giving up any of the company, getting into debt, or even doing much more than posting a video and project description on a website.

Underground VC

Kickstarter solves three huge problems for entrepreneurs. First, it simply moves revenues forward in time, to right when they're needed. One of the reasons startups traditionally have to raise money at the start is to pay for product development, tooling, purchasing components, and manufacturing, all of which they'll presumably get back later when they sell the products. But if they can turn those sales into *presales*, which is essentially what Kickstarter does, they'll have the money when they need it and won't have to raise venture capital or take out a loan.

Second, Kickstarter turns customers into a community. By backing a project, you're doing more than pre-buying a product. You're also betting on a team, and in turn they update you with progress reports and respond to suggestions in comments and discussion forums during the product's genesis. This encourages a sense of participation in the project and turns backers into word-of-mouth evangelists, which helps projects go viral.

Finally, Kickstarter provides perhaps the most important service a new company needs: market research. If your project doesn't hit its funding target, it probably would have failed in the marketplace anyway. Getting that information before you've made the time and investment in developing and manufacturing the product is invaluable and "de-risks" one of the most hard-to-access factors in any startup.

All this makes perfect sense, but it just wasn't possible before the Web. What this sort of crowdfunding offers is simple: a way for the people who most want a product to help make it happen. Each pays no more (and usually less) than they would pay anyway when the product comes out, but simply by paying earlier and delaying receipt, they collectively remove one of the greatest barriers to small-business innovation: early-stage capital.

What's more, the Web helps find these people, wherever they may be. How would you know what the market was for a jellyfish tank before the Web? Is it people who already have a fish tank? People who have lava lamps? People who like kinetic art? None of the above, but rather an entirely new class of consumers who just happen to love the idea of jellyfish on their desks—but only once presented with that idea? How would you even find out? And how much would it cost to do so?

Kickstarter and markets like it let such people find you. It's the ultimate social capital. Word of mouth will send news of a project to the most receptive people through paths that are often totally unpredictable. The means of transmission themselves are prosaic: e-mail, Twitter, Facebook, other social media. But the degrees of separation they connect are the real magic, reflecting latent knowledge about people's desires that can be identified only by the combination of the people they know and ideas that are compelling enough to pass along (what social scientists call *memetic*).

How did you come to hear of your first Kickstarter project (assuming you have)? Was it a friend who thought you might be interested?

The feed of someone whom you follow on social media? Coverage in the news in some area you follow?

The point is that you probably didn't go to Kickstarter looking for it. It found you. And if you responded, you were the right target audience even though nobody might have been able to guess that beforehand. So Kickstarter is not just money-raising, it's market research. It surfaces demand that could often not be found any other way.

Maker vs. multinational

On April 12, 2012, Sony announced with its usual great fanfare the U.S. release of its new Smartwatch, a sexy $150 gadget that would let you read texts, e-mails, and social status updates from your wrist, thanks to a Bluetooth connection to your phone. Although this is the sort of thing that would once have made headlines—Sony takes on the wrist!—it was almost completely ignored. Why? Because a day earlier, a small startup team of engineers and hardware hackers working on the ground floor of their founder's apartment building in Palo Alto had announced their own watch on Kickstarter . . . and it was simply better.

The Kickstarter project, called Pebble, had a crisp sunlight-readable e-paper display rather than Sony's OLED color display. Although color is usually preferable for computer screens, when it comes to watches, color means dim screens in sunlight, shorter battery life, and the need to push a button or shake the watch to show the time, reminiscent of the original LED watches from the 1970s. Unlike the Sony watch, which worked only with Android phones, Pebble also worked with the iPhone, and even though the Sony watch had already been out for months in Europe, Pebble ran more apps. And it was offered at $115, nearly 25 percent less than the Sony product.

In short, a few Maker-style entrepreneurs had outdesigned, outmarketed, and outpriced one of the biggest electronics companies in the world. And then, thanks to Kickstarter, they got ready to out-sell Sony, too.

The Pebble team set a Kickstarter target of $100,000. It reached that in just two hours (I was one of those early backers). And then it kept on going. By the end of the first day, it had passed $1 million. By the end of the first week, it had broken the previous Kickstarter record of $3.34 million. After a little more than three weeks, Pebble had already passed $10 million in backing and had pre-sold 85,000 watches. At that point, the team declared the product sold out and got on a plane to Hong Kong to figure out how to actually make such a huge batch of electronics (although they had made smartwatches before, the most successful of them had sold only 1,500 units). Before Pebble's monthlong Kickstarter fund-raising period was over, it had already achieved the most successful smartwatch launch of all time—and all before actually shipping a single watch.

What was particularly interesting about the Pebble Kickstarter phenomenon was how the design team responded to the crowd of customers. First the backers asked for better water-resistance, so the Pebble team figured out how to make the watch waterproof so you could swim with it. Then they asked for Bluetooth 4.0, with its lower power consumption, rather than the original Bluetooth 2.0 (or Sony's 3.0). So the team, emboldened by its flood of orders, went looking for the right 4.0 modules and were able to source them, giving the watch better battery life and making it more future-proof. Finally, other Kickstarter projects joined the parade and announced that they would be writing apps to run on Pebble, including Twine, an "Internet of Things" device that could let Pebble do things like tell you when someone's knocking at your door.

As of this writing, Pebble has not yet shipped its watches (they're due in September 2012), and perhaps production glitches will mar or delay the launch. But even before that, it's not hard to see in Pebble a superior model: a small team using crowdfunding to move more quickly in all ways—R&D, finance, and marketing—than a lumbering electronics giant. To be sure, these were not rank amateurs making their first product; the Pebble team had been together for three years and had already raised seed funding and shipped a smartwatch for the BlackBerry phone (which didn't do very well). But they were still a startup, with twentysomething founders figuring it out as they went, and prototyping with 3-D printers and Arduino open-source processor boards, just like so many other Makers. What Kickstarter did was catapult them from just another small company trying to get a break to an overnight viral hit—with money attached.

The future of funding?

Today, crowdfunding is big and getting bigger fast, and is attracting notice from Wall Street to the White House. The next step in crowdfunding is to go from simply making a donation or preordering a product to actually investing in the company itself. But such investment is heavily regulated by the Securities and Exchange Commission (notionally to protect small investors) and is typically limited to accredited professional investors.

However, as Paul Spinrad pointed out in an O'Reilly analysis of the issue:

> These laws were enacted to protect unsophisticated investors from fraud, but they also prevent people from investing in small businesses in their own neighborhoods, or garage ventures launched out of communities of interest that they belong to—despite the likelihood that their personal ties to such investments give them a better basis for evaluating risk (and contributing to success) than some mass of SEC filings cooked up in an office somewhere. And so, in the name of investor protection, the investments industry currently has a monopoly on all the invested assets of the non-millionaire public. People can't invest in the people they know from their own communities; they can only entrust their money to the choices contained in a managed menu of exclusively non-local, large-scale investment products.[42]

A number of entrepreneurs, technology leaders, and even celebrities such as Whoopi Goldberg petitioned Congress to rethink this, and carve out a way for individuals to invest small amounts (less than $10,000, or 10 percent of the investor's income from the past year) in companies that they believe in.

Washington listened. In April 2012, crowdfunding became part of President Obama's Jumpstart Our Business Startups (JOBS) Act, which he signed into law. The act makes it easier for small companies to use regulated Web-based crowdfunding sites such as RocketHub, Crowdfunder, and Launcht to raise up to $1 million in investment money from regular people, not just qualified Wall Street investors, without the laborious accounting and public disclosure rules of a traditional stock market listing.[43] Although some are concerned that such equity-based funding (as opposed to the simple preordering and charitable backing of Kickstarter) could lead to fraud, the hope is by having the SEC regulating the websites rather than the companies, they can help self-regulate the industry. And because the total amounts are small, the capacity for damage is limited.

The point is to unlock an economic engine that can drive innovation, even as the traditional financial industry pulls back. As Dominic Basulto put it in a *Washington Post* essay, "There is a unique, underground venture capital economy happening right now in America that is, in many ways, off the radar screens of economists. When we tally up the economic indicators, the conventional wisdom seems to be that economic growth in this country has stalled. Yet, that same conventional wisdom ignores the economic activity on DIY sites like Kickstarter."[44]

Social capital

Let's return to the jellyfish example to see what makes this model so powerful. Consider the advantages Andon had by going the Kickstarter route with his project, rather than to a bank or traditional investor:

1. He raised the money without having to pay interest or give up a portion of the company.

2. The process of raising the money also served as free market testing. If he hadn't been able to hit his target, he probably also wouldn't have been able to sell the tank. Raising money directly from your future customers improves the chances that you'll be successful once the product hits the market.

3. The public fund-raising effort got attention from everything from popular blogs to NBC television, serving as free marketing. Grassroots funding leads to word-of-mouth support.

Crowdfunding is venture capital for the Maker Movement. Just as the tools of production have been democratized, creating a new class of producers, so have the tools of capital-raising, creating a new class of investors. Not investors in a company but in a product or, to be precise, in the *idea of a product*. And not investors who expect a financial return, but rather investors who expect to be repaid in kind with the product itself, whether by actually getting it (because they donated enough) or by experiencing the emotional reward of knowing that they had something to do with bringing that product into existence.

The act of "making in public," which is what Kickstarter project leaders do, *turns product development into marketing*. The creator posts an idea, then updates frequently on the progress to completion. Backers comment and the creator responds, evolving the product in response to feedback. In the course of this public exchange, money is raised, but, more important, a product develops a constituency. The backers are rooting for the product not only because they've put some money into it, but also because they feel a sense of co-ownership in its creation. Making in public is an incredibly effective form of advertising, but rather than having to pay for the promotion, you can actually get paid instead.

Beat that, Madison Avenue.

What's more, it's fun. As Sarah Dopp explains at Culture Conductor, a Web community blog:

Most of Kickstarter's magic mojo is simply that they made a game out of raising money. Here are the rules to that game:

1. Set a deadline. Let people know there is a limited time to this campaign.

2. Set a minimum funding goal. "If we don't reach this number, the project won't have enough funding to happen."

3. Enforce the deadline and the funding goal. The campaign STOPS at the deadline, and if you didn't meet the goal, the project doesn't happen. (This is where Kickstarter is most valuable: they play bad cop about the rules of the game, while you get to play good cop and try to get people excited.)

4. Set up tiered levels of giving, and promise people different thank-you gifts for each level.

5. Let the fund-raisers keep full ownership of their projects. (It's not investment; it's sponsorship. It's pre-selling. It's generosity.)[45]

This is not without risk, of course. There is no guarantee that the entrepreneur will actually make the product or that it will be as good as was promised. Nor is there a promise of how long it will take. And if the entrepreneur drops the ball or simply disappears, there is no easy mechanism for the donors to get their money back. Technically, you are making a donation to a cause. Although you have been promised that you will get a product in return, there is no binding legal agreement to ensure that.

Kickstarter, like many of the sites like it (IndeGoGo, RocketHub, and Funded By Me, to name just a few), counts on transparency and the sophistication of Web users to assess risk themselves to protect against fraud or incompetence. It advises backers to use their own judgment, but offers no protections of its own. Its advice to prospective backers:

> Each project is crafted solely by its creator, and it's up to them to make the case that they can successfully bring their project to life. Part of every creator's job is earning their backers' trust, especially backers who don't personally know them.
>
> The Web is an excellent resource for learning about someone's prior experience. If someone has no demonstrable prior history of doing something like their project or is unwilling to share information, backers should consider that when weighing a pledge. If something sounds too good to be true, it very well may be.

If Kickstarter were helping companies raise funding or attract investors, it would be regulated by the Securities and Exchange Commission, and all sorts of rules and protections would kick in. But it is not. It is simply giving people an opportunity to contribute to a cause, and in this case the cause is the creation of a product they want. You're not even backing a company; you're just backing a specific project.

It's a clever way around many of the barriers that keep most small companies and inventors from raising enough money to get started. No single person puts up more money than he or she can afford, and typically people only back products that they personally want and understand.

No doubt there will be some disasters ahead. Most likely are the naïve inventors with a good idea but absolutely no competence in manufacturing who discover that they have badly underpriced their product and are unable to make it for the promised cost. Teams may fall apart, personal issues may arise, and some people will just flake out. And then, inevitably, there will be fraudsters. But so far the social support and accountability that come with transparency have prevented the usual catastrophes. And the service is growing at an astounding rate.

As of May 2012, three years after its founding, more than 47,000 projects had been launched on Kickstarter, of which more than 40 percent were successful, raising a total of $175 million.[46] Most were just a few thousand dollars for music, film, and other arts projects (for which Kickstarter was originally intended), but there were also hundreds of successful physical products. Two dozen such projects, like the Jellyfish Tank, raised more than $100,000 each.

Other examples include Scott Wilson, a former creative director at Nike. With his connections, he didn't need crowdfunding for his idea for a special strap that could turn an iPod Nano into a wristwatch. But he chose that route anyway because he wanted the direct feedback and simplicity of the Kickstarter process. His TikTok+LunaTik proposal raised nearly a million dollars. Sixty days after his Kickstarter fund-raising period closed in December 2010, Wilson shipped more than twenty thousand of the watch cases.

What Wilson avoided by going this route was the prosaic path of corporate product development: layers and layers of approval processes, which tend to favor the conventionally tried and true over real innovation. As Carlye Adler put it in *Wired*:

> Build a better mousetrap and the world is supposed to beat a path to your door. It's a lovely thought, one that has inspired generations of American inventors. Reality, though, has fallen somewhat short of this promise: Build a better mousetrap and, if you're extremely lucky, some corporation will take a look at it, send it through dozens of committees, tweak the design to make it cheaper to manufacture, and let the marketing team decide whether it can be priced to return a profit. By the time your mousetrap makes it to store shelves, it is likely to have been fine-tuned and compromised beyond recognition.[47]

Take Peter Dering, a civil engineer and an expectant father with an idea for a device called Capture that would allow you to easily clip a camera to your clothes or backpack. He, too, could have pitched his idea to a camera accessory company. Instead, he decided to go it alone. His Kickstarter project raised $365,000 from more than five thousand backers. This, he wrote, "transformed my life. On May 2, 2011, I launched Capture as a guy with a dream, out on a limb. Seventy-five incredible days later, I am a father with a business."

An open-source flashlight raised $260,000. A stainless-steel pen raised $282,000. A camping hammock raised $209,000. And so on, for hundreds more (I myself have backed everything from a three-string guitar kit for kids to a desktop CNC machine). Kickstarter has become the favored funding route for inventors everywhere, or at least for those who can put together a video and story that describes their vision in a way that compels people to buy into it.

The accidental bank

Kickstarter's origins go back long before its founding in 2009. One of its cofounders, Perry Chen, was living in the French Quarter of New Orleans in 2002, working on his own electronic music and dreaming of hosting a great DJ show with the Austrian DJs Kruder and Dorfmeister. Trouble was, it would cost $15,000 up front. And although Kruder and Dorfmeister are huge in the DJ scene now, they weren't then. What if nobody showed up? Chen would be wiped out.

The risk scared him off from doing that concert, but he continued thinking about the problem. Anything sufficiently new is risky, but the number of people who have the resources to handle large financial risk is small. What if you could just charge people up front (not such a radical idea, given that's the way most concert sales work) and, more important, not have to commit to the show if the sales weren't strong enough? That way the organizer wouldn't have to put up any money and the bands would go only where they were sufficiently wanted.

A few years later, Chen moved to Brooklyn and was waiting tables at a hipster diner called, predictably, "Diner." There he struck up a conversation with Yancey Strickler, a brunch regular, and started to tell him about his idea. Before the Web went mainstream, the idea of surfacing demand and pre-funding projects was theoretically clever but impractical. But now it might just be worth trying. Strickler loved it (Chen told Adler it was "the best idea any waiter pitched him that year"), and the two decided to build a website to try it out.

Today, Kickstarter is a multimillion-dollar Web company that's trying as hard as it can to stick to its indie roots. Its building at 155 Rivington Street on Manhattan's Lower East Side is nothing to look at. The only sign on the front, painted in gold letters, reads UNDER-WEAR (a former tenant). Inside, it looks like the entry to a group house.

Chen and Strickler are still slightly uncomfortable with the rise of Kickstarter as a financing engine for physical goods. They had originally intended it mostly for the kind of music and film projects that the record labels and Hollywood weren't willing to take a chance on, along with art, theater, comic books, and fashion. But at the core was bankrolling creativity, and more and more creatives were getting interested in making physical goods. It was just too hard to draw the line, so they didn't. A team of twenty-five approves projects before they are listed, mostly based on the quality of the presentations, not on their artistic merits. The biggest projects on Kickstarter, like it or not, are consumer goods. It simply fills a market need that was there all along, waiting for someone to tap it.

Voting capital

For all Kickstarter's egalitarian charms, once projects are funded, creators are pretty much on their own to get them actually made. As they'll quickly discover, the idea is the easy part. Supply-chain management and manufacturing are much harder, to say nothing of just running a small business. What if a community could help decide which user-submitted product ideas get made, just like Kickstarter, but then a team of product-development professionals helps steer the project, handling all the tricky factory issues?

That, in a nutshell, is the model of Quirky, which launched around the same time as Kickstarter in 2009 and is growing just as fast.

Ben Kauffman, its founder (age twenty-four at the time of this writing), got his start in his senior year of high school, when he somehow persuaded his parents to take out a second mortgage on their house to fund the creation of mophie, a company to design and make iPod accessories. He sold that company in 2007 and his next project was to create a website where people could vote on ideas and give suggestions on how to improve them. Although that never took off as a stand-alone site, it became the technological foundations of Quirky, which sought to combine the two ideas: using the crowd to develop better products like, well, iPod accessories.

Today, to be fair, Quirky does a lot more than that. Every week it puts into production two new products invented by its community. They tend to be handy "solution" household accessories, like expanding towel racks and closet organizers, mostly under fifty dollars. Quirky has a display rack in Bed Bath & Beyond, a big American home-products retailer. None of this is world-changing stuff, but the products tend to be well designed, attractive, and actually useful. It's hard to peruse the list and not find something you wouldn't mind having.

As I write this, the hot product at the moment at Quirky is the "Pivot Power" flexible power strip. It's like a regular power strip, but each outlet can pivot so that bulky power adapters don't block neighboring outlets. Designed by Jake Zein, a software programmer from Milwaukee, Wisconsin, it's classic Quirky: clever, clearly solving a problem, stylishly designed, and slightly inessential. It's the kind of thing you'd see in the store and think, "Yeah, I hate it when I can't fit power adapters in all the power strip outlets," admire its design, and maybe buy one. You don't need it, but once you've seen it you might want it.

This is not an accident. Quirky's products are the result of a remarkable series of public review steps, each of which weeds out bad ideas and improves good ones. Hundreds of people have had a hand in every Quirky product, either originating the idea, suggesting some change, or voting on which variation they prefer. Amazingly, they all get paid, ranging from the person with the original concept to everyone else who had "influence" in a final product, even just voting on a winning design.

For most of them it's just pennies, but the original inventor can earn thousands of dollars. Not millions, but not nothing. And they don't have to do much work—just describing the idea and submitting some sketches. In all, 30 percent of the sales from Quirky.com and 10 percent from retail partners goes to the community. Of that, 35 percent goes to the inventor; the rest goes to others who helped improve or select the winning design.

The way it works is this:

- Anybody can submit an idea, but it costs ten dollars to do so. This is just to keep out spammers and cranks.
- Community members vote for ideas they like, and offer comments.
- The most popular ideas go to the next phase, design. Both the inventor and Quirky's own professionals submit designs. The most popular design wins.
- More voting ("influencing") happens for the product name, tagline, feature set, and other branding.
- Quirky's engineers make the winning design manufacturable and work with a factory to make it.

As at Kickstarter, there are countdown clocks and competitions everywhere—the whole thing feels like a game. You don't need to have any ideas of your own to participate and feel as though you're helping create things, or at least improve them. And it suits everyone from words people (names and taglines) to visual thinkers (design). Top influencers participate in dozens of projects and can earn thousands of dollars. It can be addictive, they report. Partly it's the act of improving ideas, but equally it's the gamble that the product you vote for will ultimately be made and become a big hit.

At the core, what the Quirky community represents is crowd-sourced market research. By getting so much feedback at each stage of the process, Quirky reduces its risk. The products that get the most votes are the ones most likely to get the most sales. That way Quirky can put its own engineers and designers to work only on the products that are most worthy of their time. Like Kickstarter, Quirky also uses a presale process, where products are made only if they reach a certain number of purchase commitments. (Your credit card is charged only if the product goes into production.)

This is Making for people who don't actually want to get their hands dirty. They can participate in every step of conjuring a product into existence, but they don't have to make the prototype themselves. All the initial physical work is done in Quirky's offices, which are outfitted with a high-end 3-D printer and a full suite of other digital prototyping tools, and the manufacturing is done by Quirky's factory partners, mostly in China. The community can influence the final product, but they can't completely control it. At the end, they're just helping a professional design team work faster and better. And in turn, the design team cuts them in on a piece of the action, in both money and glory.

Industrialized crafting

Finally, at the other end of the spectrum is Etsy. This is by far the biggest of the three Maker markets I've profiled here. Launched in 2005, it now has more than 15 million members and did a half billion dollars' worth of sales in 2011. As of April 2012, it had 300 employees and was selling $65 million in goods per month from 875,000 sellers to 40 million visitors from around the world.[48] Indeed, at an estimated value of $688 million after six years, its growth is spookily similar to eBay's in the early 1990s—a fast-growing marketplace for the Long Tail of things.

What is being sold? Handmade goods. That's right. So far, Etsy has just been arts and crafts on an epic scale. The range is incredible, from fine art to needlepoint, with a lot of jewelry and hipster ephemera in between. Each product is made by someone (the rule at Etsy is that everything must be handmade in some way, although that doesn't mean production equipment can't be used as well).

I've bought everything from cool panda stickers for my daughters' MacBook to some awesome silk-screened prints of scientists' names and symbolic images done in the style of rock-band posters, which are now on the wall of my workshop. (Tesla and Bohr are my favorites.) All around my office people have Etsy purchases: jewelry, bookends, furniture, clothes. It plays into the generational quest for individuality and authenticity—real stuff from real people, not packaged culture from companies. Etsy stuff is sometimes gorgeous and sometimes just odd (there is a whole site, called Regretsy, focused on the most comic of the head-scratchers available there), but it is always unique. If what you want is something created by a person and not just by a machine, Etsy is a gold mine.

Unlike Kickstarter and Quirky, Etsy doesn't try to help Makers fund or create their products. Instead, it's simply a way to sell them, with a strong social component that comes from its focus on handmade goods and the crafting and arts communities that make them. Like eBay, Etsy offers simple ways for sellers to create their own listings and handles the payment processing. It charges twenty cents for each listing for four months, and 3.5 percent of the sale.

There is some debate as to whether Etsy really is a viable marketplace for small businesses. Its emphasis on handcrafting means that sellers are generally prohibited from scaling up with more-efficient automated production techniques or outsourcing some of the work. It's hard to get noticed in such a huge marketplace, and the listing fees necessary to show up in a search can add up. And all that competition can drive down prices.

Although some Etsy sellers make a living at it, most do not, and tales of that grim moment when a seller actually calculates how much she or he is making per hour (and how poorly that compares to flipping burgers at McDonald's) abound; suffice it to say that it's not about the money for most of them. For most, this is a hobby or their art, and much of the incentive lies in finding an audience for that, even if not much money follows. But for the others, who do want to make a go of it as a business, Etsy may be a place to start, but it's not the platform for growth. For that, they need to build their own companies and learn how to do real manufacturing, the twenty-first-century way.

Fortunately, Etsy is now moving in that direction, too. Although it intends to remain a place for crafters, it also intends to become a place for entrepreneurs who are using Maker-style manufacturing to grow their businesses. The handmade rule will give way to hand-designed and perhaps machine-made, or even outsourced manufacturing (the rules are being developed as I'm writing this). The point is to catalyze a new kind of cottage industry, one that really could become an engine of a new micro-manufacturing economy.

As Chad Dickerson, Etsy's CEO, put it in the company's first small-business conference in late 2011:

Decades of an unyielding focus on economic growth and a corporate mentality have left us ever more disconnected with nature, our communities, and the people and processes behind the objects in our lives. We think this is unethical, unsustainable, and unfun. However, with the rise of small businesses around the world, we feel hope and see real opportunities: Opportunities for us to measure success in new ways . . . to build local, living economies, and most importantly, to help create a more permanent future.

Right now, he noted, Etsy is still small compared to the global economy—hundreds of millions of dollars versus tens of trillions. But as Etsy expands around the world, it is bringing its model with it, from France to Germany. With its expansion comes a greater focus on growing small businesses, not just selling arts and crafts. Yet its roots remain at the human scale, with a person and face behind each product. "Don't think of it as Etsy becoming more like the rest of the world," Dickerson said. "Rather the rest of the world becoming more like Etsy."

Chapter 11

Maker Businesses

**What starts as a hobby can become
a mini-empire.**

All Makers who aspire to become entrepreneurs have heroes. These are people we read about who started with little more than a passion and access to tools, and then just didn't stop. They kept making, building, and taking chances until they had a real business. You can still see the path from the basement workbench to the marketplace, and the consequences of things having been built by hand.

This chapter is about three of my own Maker heroes. One, Burt Rutan and Scaled Composites, starts in the 1970s, at the beginnings of the modern DIY movement, and goes all the way to space today. Another, BrickArms, a Lego accessory company, is a classic Long Tail business driven by passion, some cool tools, and the Web. Finally, there is Square, one of the hottest companies in Silicon Valley, which was born when a Maker craftsman and a Web visionary got together to create the ultimate hardware/software combination, one that could someday transform the financial industry.

The ambitious hobbyist

The desert town of Mojave, California, is one of those crossroads outposts you need a good reason to visit. The wind blows hard year-round, and snakes warm themselves on the roads in the morning. A few small hotels mostly house sun-baked construction workers erecting hundreds of massive wind turbines on the nearby rocky hills. There's one bar, Mike's, where the jukebox plays heavy metal loud, and hard men with tattoos drink beer with few words. Not much else stays open after 10:00 p.m., although you can find a dogfight if you know whom to ask.

But look to the clouds above Mojave and none of this matters. Up there, in the thin desert air, can be found some of the most fantastic machines ever imagined. The Mojave Air and Space Port, Mojave's airport, is the civilian counterpart to the nearby Edwards Air Force Base, where experimental aircraft have been punching holes in the sky since World War II and the test pilots who broke the sound barrier and reached the limits of the atmosphere became the first astronauts. This is Right Stuff territory. Men still wear flight suits, and hangar doors open to reveal vehicles that seem conjured from the covers of sci-fi novels and the sketched imaginations of schoolboys.

Today Mojave is the home of many of America's commercial space companies. One of them is Scaled Composites, the aviation company founded by the legendary Burt Rutan. At the entrance to the Mojave Air and Space Port is a three-story craft called the Rotary Rocket, a Scaled design that was intended to blast off like a rocket and land like a helicopter (it actually flew a short hop once). Past it, a mile-long row of hangars hold even more ambitious vehicles designed to rekindle an adventure with the heavens that was somehow lost between Apollo and the grinding bureaucracy and cost of the Space Shuttle.

Scaled's spinoff, The Rocket Company, is now building a fleet of launch vehicles for Virgin Galactic, Richard Branson's space tourism venture that is scheduled to begin operations in late 2012. The vehicles come in a pair: SpaceShipTwo, a sleek bullet of a spaceplane with a unique tail that pops up to a 45-degree angle on descent to slow the aircraft with a controlled stall after it has taken its passengers to the edge of space, and WhiteKnightTwo, a 747-sized four-engine giant that carries SpaceShipTwo aloft, along with a cabin full of other passengers who will get a zero-G parabolic ride on the way back. Both are descended from SpaceShipOne and WhiteKnightOne, which won Scaled the Ansari X-Prize for the first commercial flight to space in 2004.

Like everything else Scaled makes, the spacecraft are constructed almost entirely of fiberglass and carbon fiber. It's a matter of some irritation to Burt Rutan, who retired in 2011, that the landing gear is still steel and aluminum; they are among the last vestiges of the metal-aircraft era that Scaled was created to end (thus the Composites in its name). Everything else is fiber, foam, and resin crafted to be stronger, lighter, smoother, and longer-lasting than metals.

Composite aircraft have other advantages over aluminum. They can take almost any shape, which is why Scaled's aircraft seem almost *grown*, not built, with graceful organic curves and slender, tapering booms. Composites are light and tough; flexible where they need to be and rigid elsewhere. And, perhaps most important in the context of this book, they can be made by almost anyone. All you need to craft a fiberglass aircraft is a foam shape on which to lay the sheets of material, a brush to spread on the resin, and a plastic sheet to hold it down while it cures, creating a smooth surface.

What makes Scaled's story so relevant to the Maker Movement is that it shows just how complex and sophisticated Maker companies and manufacturing can be. Composites, for example, are a classic Maker technology: they have democratized much of advanced aircraft manufacturing. You can lay up a wing as easily in your garage as Boeing can in its biggest factories. No special tools are required—if you've made a papier-mâché bowl, you'll understand the concept. Through the miracle of materials science, resins and threads can transform into surfaces lighter than aluminum and stronger than steel. It takes some skill to do it right, but nothing that can't be learned over a few weekends.

In fact, Scaled and Rutan got their start making composite kit planes for homebuilding hobbyists, much as kit cars also use fiberglass bodies. The same techniques that will take Virgin Galactic passengers to space began as ways to make wings and fuselages that were cheaper and easier to put together by amateurs. (Before you contemplate making one yourself, note that the average kit plane takes five thousand hours to finish, which is the equivalent of two and a half years of full-time work. Your marriage may not survive it.)

Every summer, in Oshkosh, Wisconsin, some 100,000 aviation hobbyists gather for the largest air show in the world, a festival celebrating the DIY spirit. It's run by the Experimental Aircraft Association, which is not just a community but also a Federal Aviation Administration regulatory category, which lets aircraft DIYers fly their own creations without having to go through the normal commercial certification process and flight rules. The show is a fly-in, so homebuilders from around the world fly thousands of miles in their creations to get there. There are hundreds of Rutan-designed planes, along with everything from restored World War II fighters to experimental electric-powered aircraft.

Although people come for the aerobatics and Golden Age of Flight nostalgia, the core of the event is hundreds of lectures and classes in Making. Fiberglass technique and metal machining. Painting and sanding. Foam working and aluminum bending. The list seems endless. Although the festival is about flight, it's clear that the community is about creating things. Few of the aircraft they build will spend more time in the air than they did in the workshop. Indeed, many of them will never fly at all. The creation of a beautiful machine is the real appeal for many.

This tinkerer DNA remains at the core of Scaled Composites. Many of the engineers rent space in the smaller hangars that line the runway in Mojave for their "projects," which are usually gorgeous small aircraft, from single-pilot pylon racers that can fly 500 miles per hour to half-size replicas of military aircraft from World War II. Others are pushing the innovation envelope, such as a team building an electric-powered single-pilot aircraft that they hope will set an endurance record for that class.

The Scaled engineers use the same techniques in their personal workshops as they use in their day jobs. First they design the aircraft in CAD programs onscreen. Then they either hand-carve huge foam blocks to form the shapes of the aircraft parts, or they send them to Scaled's warehouse-sized CNC machine to carve them by machine. Finally, they lay fiberglass and carbon-fiber sheets over the foam and brush resins over them to harden into sheets.

By day, they make spaceships; by night, they apply their skills to their more personal dream machines. The path from hobby to industry that created Scaled in the first place remains central to its culture; scratch any Scaled engineer and you'll find a hobbyist; walk just a hundred yards from their factory and you'll find their garages.

Hobby side projects are how Scaled engineers typically advance. To become an aircraft project leader, you must have proven the ability to run an aircraft project. How do you do this the first time? By doing it yourself. Scaled engineers win the respect of their peers with their homebrew builds; constructing and flying a machine of your own design counts for more than any academic degree in winning the trust and confidence of your peers. Each of the rented hangars holds not just an avocation, but also a résumé-builder, a laboratory for new ideas and a test bed for new techniques. Maintaining the link to the garage is how Scaled Composites stays ahead.

The DIY culture of Scaled Composites comes from Rutan himself. Born in 1943, his teenage years were full of self-designed model airplanes and competition victories. He figured out how to get a model plane to do a "power stall"—essentially hovering in midair by hanging on its propeller while he remotely controlled the throttle to keep it there. With this trick, he was unbeatable, able to do spot landings on model aircraft carriers and win "slowest flight" competitions with ease, as his airplane hung in the air while the seconds ticked by and the judges scratched their heads over how to handle this kid with his engineering hacks.

After a stint working in the aerospace industry on the Vietnam War–era F4 Phantom jet and some experimental hovering aircraft, he found himself drawn to the possibility that amateurs could build and fly high-performance aircraft, too. Supersonic flight had changed the shape of modern aircraft, but most civilian planes were docile and slow-flying designs that had hardly changed since the golden age of civil aviation between the World Wars. Rutan was taken by the designs of delta-wing jet fighters with "canard" wings in front rather than the usual horizontal stabilizer in the tail. The advantage of such canards is that they were designed to stall before the main wing; if the aircraft was flying too slowly or with its nose pitched up too high, the canard would lose lift first, dropping the nose and returning the aircraft to controlled flight.

Rutan launched the Rutan Aircraft Factory (RAF) and designed a series of groundbreaking amateur aircraft, starting with the VariViggen (inspired by the Swedish Viggen jet fighter) and leading to a series of Vari-Eze homebuilts that revolutionized the civilian aircraft industry with their composite materials and relatively simply construction. His designs were easy to build, fast and efficient to fly, safe, and reliable. Plus they looked incredibly cool. If the golden age of civil aviation was the barnstorming years before World War II, the golden age of the DIY aviation movement was in the late 1970s and early 1980s, when Rutan's designs brought advanced materials and aerodynamics within the reach of anyone.

Eventually, however, the economics of the DIY market proved too daunting and Rutan shut down RAF, instead focusing on Scaled Composites, the company he had started to design aircraft for commercial and military customers. The problem with the homebuilt market of the time was that companies tended to sell plans, not kits. The plans could cost as little as twenty-five dollars, but led to years of customer-support expectations from homebuilders with questions and requests for help. It was, in short, a terrible business.

Even when companies switched to selling kits instead, they ended up with all the aerospace challenges of tooling, component sourcing, and legal liability, but rather than selling hundreds of aircraft for millions of dollars each, they were selling dozens for a few tens of thousands each. It's a tiny market with huge risks. Rutan's most popular homebuilt, the Vari-Eze, sold fewer than eight hundred units in its entire life. A single Scaled Composite commercial customer could offer more profit with infinitely less hassle. As much as Rutan's roots were in the DIY movement, the economics of developing advanced designs in secret for big companies and government contracts were irresistible. Most of all, Rutan wanted to design groundbreaking aircraft, not feed the endless demands of the kit business.

Today Scaled Composites is owned by Northrop Grumman. For every high-profile design like SpaceShipOne, there is a cruise missile prototype or stealthy drone for the defense industry. The DIY roots are still there in all the side projects of the Scaled engineers in their personal hangars along the flight line at the Mojave airport. But the company itself is a high-security operation.

Rutan's career is an object lesson in both the potential and the limits of the Maker Movement. He used the democratized technology of composites to bring advanced aerospace concepts to amateurs. But the barriers to entry in manned flight, from the costs of manufacturing to the risk of lawsuits, turned out to be still too high to create a viable challenge to the existing industrial aerospace model.

Because the aircraft carry humans, they must go through endless regulatory and legal review, at great cost of money and time. That's still something only big aerospace companies can afford, which is why Scaled is now owned by one. But Rutan himself, now rich and retired, is one happy Maker.

The Long Tail of Lego

Turn back the clock to Rutan's origins as an enthusiast industrializing his hobby, and you've got Will Chapman today. Chapman has three sons who, like many, were obsessed with Lego until about the age of eight. Then, like a lot of boys, they started playing with toy soldiers, and Lego couldn't keep up.

Lego, as a family-oriented company, has some rules about guns.

With few exceptions, it doesn't make twentieth-century weapons. You can go farther back into history and have Lego swords and Lego catapults, but not Lego M-16 automatic rifles or rocket-propelled grenade launchers from today. Or you can go forward into fantasy and have Lego laser blasters and plasma cannons, but you can't have World War II machine guns and bazookas.

That's a perfectly fine policy for Lego, but the consequence is that it tends to lose its customers around the age of ten, when they go through their war phase. That included Chapman's sons. In 2006 his youngest one wanted to replicate a World War II battle in Lego and was disappointed that he couldn't do it with the Lego figures he already had.

That would have been the end of the story, but Chapman is a Maker. In his Redmond, Washington, basement he has a small CNC mill and he knows how to use 3-D CAD software. So he started designing some Lego-sized modern guns. And because he could, he actually fabricated them.

To do so, he first sent the files to his desktop CNC machine, a Taig 2018 mill that costs less than $1,000, to grind the mold halves out of aircraft-grade aluminum blocks. Then he put the molds in his hand-pressed injection-molding machine, which uses regular propane like that for a backyard barbecue to melt plastic, and a lever like a water pump to force it into the mold. For the plastic he just used spare Lego blocks, to use the same ABS plastic as the real thing.

After some experimentation and revisions, he had some pretty good-looking prototypes, including an M1 infantry rifle and a sniper rifle. His son was impressed, and so he made a few more and started sharing them with other "adult fans of Lego." They started clamoring for more, and so he launched a website to sell them.

Today, his company, BrickArms, goes where the Danish toy giant fears to tread: hardcore weaponry, from Lego-scale AK-47s to frag grenades that look like they came straight out of Halo 3. The parts are more complex than the average Lego component, but they're manufactured to an equal quality and sold online to thousands of

Lego fans, both kids and adults, who want to create cooler scenes than the standard kits allow.

Lego operates on an industrial scale, with a team of designers working in a highly secure campus in Billund, Denmark. Engineers model prototypes and have them fabricated in dedicated machine shops. Then, once they meet approval, they're manufactured in large injection-molding plants. Parts are created for kits, and those kits have to be play-tested, priced for mass retail, and shipped and inventoried months in advance of their sale at Target or Wal-Mart. The only parts that make it out of this process are those that will sell in the millions.

Chapman works at a different scale. He continues to design the weapons in CAD software and prototype with his desktop fabrication tools. Once they look good, he sends the file to a local toolmaker to reproduce the mold out of stainless steel, and then to a U.S.-based injection-molding company to make batches of a few thousand.

Why not have the parts made in China? He could, he says, but the result would be "molds that take much longer to produce, with slow communication times and plastic that is subpar" (read: cheap). Furthermore, he says, "if your molds are in China, who knows what happens to them when you're not using them? They could be run in secret to produce parts sold in secondary markets that you would not even know existed." Chapman's three sons package the parts, which he sells direct. Today, BrickArms also has resellers in the UK, Australia, Sweden, Canada, and Germany. The business grew so big that in 2008 he left his seventeen-year career as a software engineer; he now comfortably supports his family of five solely on Lego weapons sales. "I bring in more revenue on a slow BrickArms day than I ever did working as a software engineer."

How does Lego feel about this? Pretty good, actually. BrickArms and the many small companies like it, such as BrickForge and Brickstix, that make everything from custom Lego-sized characters to stickers that allow you to customize official Lego minifigs, represent a *complementary ecosystem* around the Danish giant.

They solve two problems for Lego: First, they make products that wouldn't sell in large enough quantities for full Lego production, but nevertheless are wanted by Lego's most discriminating customers. This is the Long Tail of Lego, and such niche demand is as real in plastic building toys as it is in music and movies. The entrepreneurs orbiting around the Lego mother ship collectively fill in the gaps in the market, allowing Lego to continue focusing on the blockbusters its scale requires.

Second, by offering products that are particularly prized by older children, companies such as BrickArms keep them in the Lego world a few years longer, from around eight or ten to perhaps twelve. This increases the chance that they will graduate from casual play to true Lego obsession, maybe even maintaining that into adulthood (don't laugh—Lego's "Architecture" series of famous building kits is sold in bookstores and museum shops for around $100 each). If so, they may become the buyers of Lego's most elaborate kits, including a Star Wars Death Star and Star Destroyer, which both have more than three thousand pieces and cost $400.

So Lego by and large turns a blind eye to this swarm of Lego fan-created businesses around it, as long as they don't violate Lego's trademarks and include cautions about keeping pointy or easy-to-swallow toys away from young children. Indeed, Lego has even issued informal guidance on using the best plastics that are non-toxic and including holes in parts that could be a choking hazard, to allow for air passage.

What BrickArms and its kin represent are examples of Maker business targeting niche markets, which are often underserved by traditional mass manufacturing.

One of the triumphs of the twentieth-century manufacturing model was that it was optimized for scale. But this was also, at least from a twenty-first-century perspective, a liability. Henry Ford's powerful mass-production methods of standardized interchangeable parts, assembly lines, and routinized jobs created unbeatable economics and brought high-quality goods to the common consumer. But they were also tyrannical—"any color you want as long as it's black"—and inflexible. The price differences between small-batch and big-batch products were so great that most buyers could have either affordable products or wide choice, but not both—cheap, mass-produced products beat variety every time.

Meanwhile, the long tooling cycles of mass production meant that products had to be designed years in advance of sale, and the cost of innovation rose as the consequences of failed experimentation at mass scale rose (witness the Edsel, a radical car that set back innovation at Ford for decades). Today the same is true: the local furniture maker can compete with IKEA only by serving the rich. All those Billy bookcases out there (and I've got my share) are the marketplace saying that they don't care enough about differentiated shelving to pay more for it.

A more pernicious cost of the triumph of mass production was the decline of small-scale manufacturing. Just as in retail, where the local specialty retailer was driven out by Wal-Mart, in manufacturing, scores of car companies were overwhelmed by Detroit's Big Five (or subsumed into them) in the first half of the twentieth century. So too in textiles, ceramics, metalware, sporting equipment, and countless other industries. All succumbed to the lure of labor arbitrage abroad, while wage pressure made union relations increasingly toxic at home.

To be sure, many of these smaller manufacturers lost on their merits: their products were no better than imported goods and their costs uncompetitive. But others failed because they lost their distribution channels to the few consumers who still wanted their specialized goods (or just wanted to buy American). The grinding race to the bottom of price competition at the big-box retailers made it increasingly hard to find niche goods.

Fast-forward a half century, and two things have changed. First, thanks to desktop fabrication and easy access to manufacturing capacity, anyone with an idea can start a business making real things. And second, thanks to the Web, they can sell those things globally.

The barriers against entry to entrepreneurship in physical goods are dropping like a stone.

"Markets of ten thousand" defines the successful niche strategy for products and services delivered online. That number is large enough to build a business on, but small enough to remain focused and avoid huge competition. It is the missing space in the mass-production industry, the dark matter in the marketplace—the Long Tail of stuff. It is also the opportunity for smaller, nimbler companies that have emerged from the very markets they serve, enabled by the new tools of democratized manufacturing to route around the old retail and production barriers.

Even better, some of those companies that start with niche markets may graduate to huge ones.

The ultimate combination of atoms and bits

In early 2009, if you had visited the TechShop makerspace in Menlo Park, California, south of San Francisco, you would have seen a tall, somewhat gangly guy named Jim McKelvey at a bench, fiddling with a little block of plastic. For all anyone could tell, he was just another guy trying to learn how to use a CNC machine, albeit with a particularly unimpressive little project. What no one knew was what that little block of plastic might someday do.

McKelvey, then forty-three, was a technology entrepreneur from St. Louis. In 1990 he had started an early digital publishing company called Mira, which rose in the first multimedia wave of CD-ROMs and pre-Web online data. In those heady years in the early nineties, he and his team often gathered at a local coffee shop for brainstorming sessions. One day the coffee shop owner, Marcia Dorsey, mentioned that her son Jack was interested in computers and was looking for an internship. McKelvey agreed to meet him at the Mira offices.

At the agreed time, McKelvey was head down over his keyboard in the midst of a hellish deadline when a kid tapped him on the shoulder and said, "Hi, I'm Jack. My mom said you need some help." McKelvey looked up, surprised (he'd forgotten about the appointment), and said, "Hi. Can you wait a second while I finish this?" and returned to his work.

Thirty minutes later, McKelvey realized that he had totally forgotten about the visitor. He looked up and Dorsey was, amazingly, still standing in the exact same place, with his arms straight at his sides. He apparently hadn't moved or said a word for half an hour. This was strange, even by programmer standards.

To be fair, it was equally strange for McKelvey to have forgotten he was there. (Dorsey, in his defense, has said that he was happily entertained looking over McKelvey's shoulder and trying to find the bug in his code.) But this just means they were well matched. McKelvey's own quirks are legendary, including spending three years teaching himself to play only the notoriously difficult third movement of Beethoven's "Moonlight Sonata," which today remains the sole piano piece he knows.

McKelvey liked Dorsey's intensity and hired him on the spot. Over time they developed an easy and successful relationship, two of the smartest geeks in St. Louis, one ten years older than the other. McKelvey gradually brought Dorsey out of his shell, while Dorsey blew everyone away with his programming prowess.

Eventually, McKelvey sold Mira and decided to turn to glassblowing, an early love and something he resolved to become really expert in (more on this shortly). Dorsey, meanwhile, moved to Oakland, California, and joined a small Web startup called Odeo, which was trying to make some inroads with podcasting software.

A year passed, Apple built its own podcasting software into iTunes, and Odeo was clearly in trouble. Its founder, Evan Williams, asked the staff whether anybody had another idea for a business. Dorsey, as it happened, did—it revolved around a concept he'd sketched out a few years earlier about instant status updates. He, fellow Odeo employee

Noah Glass, and Florian Weber, a contract programmer, hacked together a little proof-of-concept that let people broadcast SMS-style messages to people who signed up to "follow" them. They called it Twttr. Williams and the rest of the team liked it, shut down Odeo, and returned the money it had raised to its investors, and started a new company around the idea. They added the missing vowels and called it Twitter. The rest, as they say, is history.

Dorsey had finally hit the big time. But Williams was now running Twitter, and Dorsey wanted his own company. He struck up a conversation with his old boss, McKelvey, and they resolved to start a new company together. They had a few ideas about what it would do, probably involving mobile in some way. Dorsey, however, was forbidden from doing anything like Twitter as part of his non-compete agreement, and that eliminated a lot (as McKelvey dryly puts it, "there is a lot of surface area in Twitter's future"). So they went looking for another big problem to solve.

At that point, as McKelvey tells it, he was having trouble completing the sale of one of his glass pieces over the phone. A woman in Panama wanted to buy a glass bathroom faucet that cost more than $20,000, and she had only an American Express card, which McKelvey couldn't take. He had a sinking feeling that because of the limitations of the credit-card industry, he was going to lose the sale. And in that moment, he realized what he and Dorsey should do: revolutionize payments.

And that's how he found himself at TechShop, trying to put together a little plastic block. That block held a credit-card reader (which was nothing more than the magnetic head of a cassette player) that plugged into the audio jack of an iPhone. When someone swiped a card through the device, it generated an audio signal that the software in the phone could read, translate into meaningful data, and send to a website to initiate a credit-card payment. That allowed the phone to replace a bulky and expensive point-of-sale terminal. Anybody could take a credit-card payment, anywhere—they just needed a phone and this little plastic reader. The company McKelvey and

Dorsey started would be called Square, in part because of the shape of the little device.

Unlike McKelvey and Dorsey's previous companies, Square was a combination of both hardware and software: the little phone dongle was the atoms and the phone app and Web services that worked with it were the bits. That meant that they were in the electronics business, like it or not.

This was not the way Dorsey had wanted it. He was a programmer and felt sure the problem could be solved with software alone, by using the phone's camera to read the numbers on the credit card. Easier said than done. "That actually turns out to be really hard," says McKelvey. "If you don't have the card tilted just right, it's impossible to read the characters." The two fought about it, each making increasingly technical arguments about why his approach was better. There was only one way for McKelvey to settle it: "I had to build a hardware prototype to convince him that hardware was a better way."

So McKelvey went to TechShop to build a series of test credit-card readers. He'd actually started a few months earlier in the student machine shop of Washington University in St. Louis, where McKelvey teaches glassblowing. But Dorsey and Square were based in San Francisco, so to win the day he had to come to Silicon Valley and finish it there.

The first few Square devices were hand-cut. Then the next were made on TechShop's CNC machines, with McKelvey writing raw G-code script (rather than designing in a CAD program). Each version got smaller, more stylish. Dorsey was convinced—hardware it would be. The plan was to give away hundreds of thousands of the Square readers and make the money back on a cut of the transaction fees, much like a credit-card company. But that meant being able to make a huge number of the Square readers at a cost of less than a dollar each. They had to be practically unbreakable and foolproof. On the scale at which Square needed to operate, a mechanical or electrical problem with the readers would bankrupt the company.

The reason McKelvey was prototyping the devices himself at

TechShop, even though he knew little about this kind of hardware engineering, was to get firsthand experience. If the company was going to hand out millions of these gadgets, they'd better work just right. This was going to be the consumers' gateway to their service, and the physical embodiment of the company. Outsourcing the design and production processes to a contract manufacturer would have been cheaper and easier, but risky. How would they even know which design and manufacturer to choose if they didn't really understand their own product? The only way to ensure that was the case was to make the first devices themselves, to learn everything about them, inside and out.

"I hand-built fifty of those things. There's nothing like that," he says. "I know about azimuth errors and torsion errors. The knowledge of actually doing it, of having the machines under your control, is a huge multiplier. If you see it happen—see how the flash comes off the injection-molding—you realize it matters which way the head moves when you pull the injection lines to deal with the shrinkage.

"If I hadn't done it myself, that knowledge would have been intermediated. We'd have had a clunky, committee-designed product. Later, more expensive, and it wouldn't have been as cool."

When it was time to go into mass production, the loyal Missourian tried first to find a suitable injection-molding company in St. Louis, but there were none that could handle the volume and pricing. So he went to China. The final design process took place in Guangdong, with McKelvey and an engineer who didn't speak English working together until 3:00 a.m. over an outdated version of the Solidworks CAD software (the Chinese factories wouldn't use any version of Solidworks after 2007, when there was an anti-piracy crackdown). The path from Maker to Industrialist was complete.

Today, Square has a valuation in the billions of dollars and millions of customers. It has expanded from person-to-person transactions with phones to full iPad-based point-of-sale terminals, competing with such cash-register giants as NCR. Visa, the credit-card company, is an investor, in part because it sees in Square the same sort of ambition to become a global payments platform as tuned for the mobile age as Visa was for the plastic age. In the mornings Dorsey runs Twitter, where he has returned as executive chairman; and in the afternoons and very late into the evening he runs Square. Count the hours and you can see his priorities. His wealth may be more tied up in Twitter, which is worth even more billions of dollars than Square, but his heart is in reinventing payments.

Poignantly, Square's offices are in the former *San Francisco Chronicle* building, a symbol of twentieth-century industrial might. Once, huge printing presses ran day and night and fleets of trucks brought massive rolls of paper to be turned into newsprint. Now the newspaper is in decline, the presses are gone, and the space is being recolonized by Web and Maker companies. In another building in the complex, which was once used to store paper rolls, TechShop opened its San Francisco branch, and every day it's full of people just like McKelvey making what they hope will be the next big thing.

McKelvey, meanwhile, remains Square's chairman, but he spends most of his time in St. Louis. There he continues to teach and practice glassblowing. Which, as it happens, is not unrelated to his Maker moment in TechShop.

The connection is this: glassblowing went through its own Maker moment thirty years ago. Glass artistry has required the same skills for two thousand years. High and very constant temperatures are required to keep the glass at just the right malleability, which means huge furnaces with ceramic walls that hold the heat to ensure an even distribution. Glass ovens take four days to come up to temperature and can never be turned off lest you crack their walls. You have to constantly feed them with fuel. That's why, McKelvey says, there are no forests around Venice. Venetian glassmakers used all the trees.

Making glass this way has traditionally required industrial-sized operations, such as those that make Tiffany lamps today. But as is always the case with industrial-sized operations, you get only the sort of mainstream products that can support the economics of a factory. Creativity was constrained by the need to sell in large numbers.

But in the early 1960s, two glass artisans, Harvey Littleton and Dominick Labino, invented a formulation for low-temperature glass and a small propane-powered furnace that could properly melt it. Now it was possible for an individual to work glass with equipment that a small studio or community arts center could afford.

That was the equivalent of the laser printer in the PC era, or the laser cutter and 3-D printer today. Cheaper, smaller, more powerful tools means that ever more complex activities become accessible to regular people. Those 1960s innovations democratized the tools of production and a thriving glass art movement began, which McKelvey is part of today. He was the chairman of the world's largest gathering of glass artists, has written a textbook on the craft, and runs a studio, the Third Degree Glass Factory in St. Louis.

McKelvey is a classic Maker who has built a business from what was once a hobby. So when he and Dorsey decided to start Square twenty years later, those same instincts drove him to DIY the hardware. That allowed Square to get to market sooner, with a better product—they were able to refine their design and understand its strengths and weaknesses faster because they had made it themselves.

Today, Square is so successful that some of the biggest financial payment companies in the world have started running attack ads trying to encourage their customers not to switch. Verisign, which makes point-of-sale credit-card readers, thinks the Square method is less secure than its own. Its ads say, "The glassblower stole your credit card." McKelvey loves it. It reminds him of where he came from—and the perils of big companies underestimating Makers.

Chapter 12

The Factory in the Cloud

Once manufacturing went online, nothing would be the same again.

Mitch Free was destined to be a worker, and probably not much more. He grew up in Tyrone, Georgia, then a town of 160. His father ran a small construction business; Mitch helped out when he felt like it. He went to college for six weeks before he decided that remedial English wasn't for him, and quit. He then enrolled in a technical school for a one-year course and selected machining on a whim (electronics, which he was more interested in, was full). Once finished with that, without honors, he started work in a machine shop called Dixie Tool and Die, pressing a button on a stamping machine that made window linings for Ford vans. Sometimes he hand-polished metal.

It was 1982, he was twenty years old and married to his high school girlfriend, and this looked like a preview of the rest of his life.

Then, one day, his boss asked whether anyone on the shop floor knew anything about CAD/CAM design. The Ford Motor Company had given the shop a big contract and it required digital files. Free, who knew nothing about digital stuff, put up his hand anyway. Why? "I was getting really depressed about my career choices," he says. Also, nobody else wanted to do it.

He crammed with some technical manuals and went to Ford's operations in Dearborn to learn what the car company wanted. Then he started digitizing the machine shop's designs to comply. He got better at it. First he manually edited the machine-code files; then he learned how to program software to do it. As sometimes happens, learning to program flipped a switch in his head. He loved it. He had finally found his calling. In 1988, Northwest Airlines, which had a maintenance facility in Atlanta, recruited him to create digital copies of replacement airplane parts that the manufacturer couldn't supply, so the airline could fabricate them itself if needed.

Over time, he became the "innovation guy" at Northwest and grew more expert at digital tools, including a CNC machine he built that could automatically examine turbine blades, looking for flaws. He was taking old DC-10s out of mothball storage and fixing them up enough for a crew to take them to Israel, where they could be overhauled and resold to a leasing company for a profit of more than $10 million each. By the late 1990s he had become the airline's director of technical operations at a time when it was becoming clear that the difference between successful and unsuccessful airlines was all supply-chain management—using global suppliers to get the right part to the right place at the right time.

That, in turn, made him realize that there was something bigger afoot than simply running an airline efficiently—the entire process of manufacturing was being reinvented by digital technologies. He took an offer to run regional sales for a CNC machine company, and in the course of doing so, he started talking to more manufacturers. He discovered that what they needed more than anything else, even more than a new CNC machine, was the ability to talk to one another. So he started hosting lunches. Then, one day in 1999, driving back from a lunch, he heard a LendingTree.com ad on the radio: "Request your mortgage. Let lenders compete for it." He realized he should be doing the same for manufacturing.

Free bought the domain "MFG.com" for $2,000 and in 2000 started an online marketplace for manufacturing. The idea was simple: companies that wanted something made would upload their CAD files to the site, wrapped with a description of how many they wanted and any other instructions, and machine shops and other manufacturers would bid on the job, just as lenders compete for mortgages on LendingTree. Companies would build up ratings over time, and highly rated suppliers could avoid the lowest-bidder trap.

This was, to be fair, not a terribly original idea. Around that time, all sorts of other "business-to-business" marketplaces, with names such as Ariba, VerticalNet, and CommerceOne (and lots of "e" prefixes from eMetals to eTextiles), were starting up in industries from cars to plastics. Driven by the dream of "frictionless digital capitalism," as Bill Gates's book *The Road Ahead* put it at the time, they were all going to revolutionize supply-chain management. Some sought to use a reverse-auction model like eBay to drive prices down. Others were consortia of big buyers in an industry designed to gang together to achieve Wal-Mart–like purchasing power (something that allowed me to use the term *polyopsony*—a monopoly of many buyers—for the first time in *The Economist*, as far as I know, which is something that I am inexplicably proud of).

In February 2000, when MFG.com was starting, there were more than 2,500 such online B2B markets.[49] Then the market crashed, and by 2004 fewer than two hundred were left. Billions of dollars of stock market value evaporated. Part of the collapse was the familiar irrational exuberance of the time. But like many other dot-com ideas, they weren't crazy—just too early. Companies weren't set up to buy electronically; many had not even moved past the fax age by then. None of their procurement systems or accounting systems worked with the new marketplaces, forcing employees to hand-type everything. What's worse, the suppliers didn't want to participate. Why should they compete in a marketplace where the goal was to drive prices as low as possible, when they could use the buyer/supplier relationships that they had built up over decades with big customers instead?

MFG.com was one of the survivors. Because it was later to start, it hadn't been hyped to the stars. There was no failed IPO; there were no massive venture rounds. Instead, it was Free and a few employees in Atlanta building a bare-bones website from scratch with Free's own money. By starting small, without the distortions of too much money and pressure, it had time to find its path.

That path was simplicity. No auctions, reverse or otherwise. No group buying or order pooling. No "frictionless capitalism." Just a place to upload files and get quotes.

Rocket science

It worked. After the dot-com crash, business started to grow nicely, and by the mid-2000s there were thousands of requests and offers placed every day. A few of them were from a small, somewhat secret group in Kent, Washington, called Blue Origin, which wanted high-tolerance parts for what appeared to be a rocket. It was, in fact, a rocket, and Blue Origin turned out to be the stealth space company started by Amazon founder Jeff Bezos. The Blue Origin engineers were so impressed by MFG.com that they brought it to Bezos's attention, who started using the site under an assumed name to check it out.

While Bezos was secretly browsing the site, Free was negotiating to sell it to Dassault Systems, a French manufacturing technology company. Just two weeks before the deal was to close, Bezos pounced and put in a counteroffer to invest in the site and keep it in Free's hands. He tossed in another $2 million for the employees and that sealed it: MFG.com would now remain independent, with Bezos as its main investor.

Today it is the world's largest custom manufacturing marketplace. It has more than 200,000 members in fifty countries and has brokered more than $115 billion in deals so far, with an average of $3–4 billion a month today.

The deals that scroll by on any given day are typically pretty prosaic stuff—injection-molded plastic enclosures, machined metal rods, fasteners, specialty cables—but they give Free an unmatched window on the world of manufacturing today. He (and anyone else who chooses to dig through the site) can see where things are being made and by whom. He can watch the flows of fabrication, the tides of tooling. The Americans sourcing in China and those who are returning to the States. The Germans sourcing in Poland and the French sourcing in, well, anywhere but Germany. It's a fascinating glimpse into culture, economics, and globalization. Forget the rhetoric—this is the raw deal flow of what companies are actually doing every day.

What's even more interesting than what's being ordered is who's doing it. It's not just big companies ordering custom parts and molds from global machine shops, but little ones, too: bike makers and furniture shops; electrical contractors and toymakers. Twenty years ago they would have had to settle for the best the local machine shop could do (at whatever price they charged), or get on a plane and try to navigate the complexities of finding a supplier in China, complete with required introductions, language barriers, and a non-zero possibility of being robbed blind.

Now companies of any size can just upload a CAD file and let the bids come to them. They get the best pricing and best products in the world without leaving their desks. Sound familiar? That's what the first wave of e-commerce offered regular shoppers. Now we're seeing the eBay and Amazon effect play out in manufacturing, too.

Why does it work so well now, and not a decade ago? The world just caught up. Along with a Web generation coming into management at traditional companies, the digital fabrication methods that captured Free's imagination have gone mainstream. The main reason MFG.com can work today when so many B2B marketplaces failed a decade ago is that companies throughout the manufacturing supply chain now all use the same file formats, from CAD to electronics. The transaction costs of closing a deal have fallen because there's less lost in translation. Everybody speaks the same language of digital manufacturing. It's as simple as that. It just took common platforms to make the dream of hyperefficient B2B online marketplaces a reality.

This is the way all successful technological revolutions work. The

Gartner Group describes this boom-bust-boom trajectory as the "Hype Cycle" of tech-driven change. After the "Peak of Inflated Expectations," there is the "Trough of Disillusionment." Then comes the "Slope of Enlightenment," and finally the "Plateau of Productivity." We've been through the first three already. Now we're enjoying the last. By the time a business process is too boring to comment on, it's probably starting to actually work.

So while the rest of us are having our heads turned by the latest buzzy social media thing, sites like MFG.com are quietly going about their work of turbocharging the world's real economic engine, making stuff faster, cheaper, better.

Open Sesame

In 1999, while I was working in Hong Kong as *The Economist*'s Asia Business editor, one of the first people I met was a hyperkinetic wisp of a man named Jack Ma, who wanted my advice on a new Web company he was setting up. Four years earlier he had taken a trip to the United States, where he'd seen his first Web browser in action. It blew his mind, as it did for many people back in the day. When he returned to his hometown of Hangzhou, he found a dial-up number for Internet access, gathered friends around, and waited three hours for the first page to load. It was thrilling. The Web existed in China! He went on to start China Pages, which is considered China's first Internet company, and ran an early e-commerce project for China's Ministry of Foreign Trade and Economic Cooperation.

When Ma came to see me, I was struck by three things. First, he was the tiniest adult male I had ever met. Not just short but small-boned and skinny. I doubt he weighed more than eighty pounds and most of that seemed to be his head, which was probably just normal-sized but appeared huge on his frame. Second, he spoke perfect English and what weight he had seemed to be entirely brain. He was brilliant and incredibly articulate and enthusiastic about the potential of the Internet, which was not a common thing to hear from mainland Chinese nationals in those days. Finally, in part because of his role with the Trade Ministry, what he was most excited about was not the consumer side, but the Web as a way for smaller Chinese manufacturing companies to break through the language and cultural barriers to doing business directly with foreigners.

What he wanted to ask me was what I thought of the name "Ali-baba." "You know," he said, "like 'open sesame.'" I liked it, encouraged him (although I seem to recall that I had some unhelpful advice about changing the tagline), and off he went.

Today, Ma is a billionaire. The Alibaba Group, which owns some of China's biggest Internet companies, has more than 23,000 employees. Its $1.7-billion initial public offering on the Hong Kong Stock Exchange in 2007 was the biggest tech debut since Google. As I write this, he is considering buying Yahoo! Last time we met, in New York, it appeared that he had put on some weight. He may be close to a hundred pounds now.

Alibaba.com is still the core of Ma's operation. It has achieved everything he set out to do, and more. It has more than 70 million users and 10 million "storefronts," both Chinese firms and producers elsewhere. Every day millions of people do what he envisioned more than a decade ago: place manufacturing orders with factories from their desks.

While MFG.com was doing this with machine shops, Alibaba was extending the model to everything and everyone. It is like the eBay for manufacturing—anyone can order practically anything to be made for them and in any quantity. I've ordered custom electric motors for a robotic blimp from a specialty motor-maker in Dongguan; I specified the shaft length, number of windings, and wire type, and ten days later prototypes were on my doorstep for my review. I was, I have to admit, stunned. I had got a Chinese factory to work for me! What else could I do with this new-found power?

From a Maker perspective, Alibaba and sites like it are an enabling technology like no other. They have essentially opened the global supply chains to buyers of all sizes, including individuals, letting them scale prototypes into full production runs.

This is not just due to Alibaba; it's also coming from a transformation in the Chinese economy and management culture. Over the past few years, Chinese manufacturers have evolved to handle small orders more efficiently. This means that one-person enterprises can get things made in a factory the way only big companies could before.

Two trends are driving this. First, there's the maturation and increasing Web-centrism of business practices in China. Now that the Web generation is entering management, Chinese factories increasingly take orders online, communicate with customers by e-mail, and accept payment by credit card or PayPal, a consumer-friendly alternative to traditional bank transfers, letters of credit, and purchase orders. Second, the current economic crisis has driven companies to seek higher-margin custom orders to mitigate the deflationary spiral of commodity goods.

For a lens into the new world of open-access factories in China, just search Alibaba (in English), find some companies producing more or less what you're looking to make, and then use instant messaging to ask them if they can manufacture what you want. Alibaba's IM can translate between Chinese and English in real time, so each person can communicate using their native language. Typically, responses come in minutes: we can't make that; we can make that and here's how to order it; we already make something quite like that and here's what it costs.

Ma calls this "C to B"—consumer to business. It's a new avenue of trade, and one ideally suited for the micro-entrepreneur of the DIY movement. "If we can encourage companies to do more small, cross-border transactions, the profits can be higher, because they are unique, non-commodity goods," Ma says. The numbers bear this out. Over the past three years, Ma says, more than 1.1 million jobs have been created in China by companies doing e-commerce across Alibaba's platforms.

This trend is playing out in many countries, but it's happening fastest in China. One reason is the same cultural dynamism that led to the rise of *shanzhai* industries. The term *shanzhai*, which derives from the Chinese word for "bandit," usually refers to the thriving business of making knockoffs of electronic products, or as Shanzai .com (the spelling of the Chinese word in English sometimes doesn't have the second "h") more generously puts it, "a vendor, who operates a business without observing the traditional rules or practices often resulting in innovative and unusual products or business models." But those same vendors are increasingly driving the manufacturing side of the Maker revolution by being fast and flexible enough to work with micro-entrepreneurs.

Today, shanzhai manufucturers are shipping more than 250 million mobile phones a year, many of them knockoff copies of iPhones and Android models, many of which are produced in relatively small quantities of ten thousand or fewer. Variation abounds, from styling to product features, in an effort to stand out. (For instance, many shanzhai phones have two or even three SIM card slots, to accommodate consumers who use different cards for home, work, and even mistresses.)

What's interesting about shanzhai is how similar the organization structures of piracy end up looking to those of open source. Once ideas and technology get into the wild, whether dragged there by pirates or placed there by developers who believe in open source, they tend to stimulate the same sort of collaborative innovation. Ideas, once shared, tend to be shared further. People who are sharing ideas tend to work together for mutual benefit. Without secrets, prices fall and accountability rises.

In a conversation with the Institute for the Future, David Li, founder of Xinchejian, China's first formal hacker space, explained why the shanzhai model is a model for open innovation, micro-manufacturing, and the future of personal manufacturing:

Shanzhai manufacturers started without much regard to the IP [intellectual property] of the original holders and share the information among themselves openly. None of the vendors participating in the ecosystem are big and there is no centralized giant among them to coordinate the ecosystem. Each one of them pulls and pushes each other to produce an efficient micro-manufacturing ecosystem that can respond to the market fast with very little overhead.[50]

As Li describes it, these companies fit neatly into the Institute for the Future's model for "lightweight innovation."[51]

1. **Network your organizations:** "The bike vendors in Chongqing hang out in tea houses and shanzhai vendors in Shenzhen have a vast network centered in the large electronics malls."

2. **Reward solution seekers:** "Penny-a-unit profits force the shanzhai collaborations to be totally solutions-driven. They don't make money if they don't deliver. 'Not invented here' is never a problem."

3. **Err on the side of openness:** "The wild west of shanzhai is all about openness. Trade secrets of big companies are flowing freely. Everything is 'open sourced' by default. If we take the [intellectual property rights] issue aside, it's really the ultimate openness we in the open-source world are looking for."

4. **Engage actively:** "The shanzhai vendors used to produce knockoffs after original vendors had the products on the market. But in the past year I have seen a lot of them act on the latest Web rumor, especially those related to Apple. It was kind of funny that there were several large-size iPhones (seven-inch and ten-inch) being produced by the shanzhai simply on the rumor that the iPad would look like a large iPhone."

The rise of shanzhai business practices "suggests a new approach to economic recovery as well, one based on small companies well networked with each other," observes Tom Igoe, a core developer of the open-source Arduino computing platform. "What happens when that approach hits the manufacturing world? We're about to find out."

The DIY factory

Finally, there is a third group of "factories in the cloud": the Web-based service bureaus that do with digital fabrication tools such as laser cutters and 3-D printers what photo services such as Shutterfly do with your pictures: you simply upload files and get back fabricated objects. They give you access to high-quality production without your having to own the tools yourself.

Perhaps the best known of these are Ponoko and Shapeways. Ponoko (on whose advisory board I sit as an unpaid volunteer) started in New Zealand as a laser-cutting service, but is now global and offers laser cutting, 3-D printing, and CNC cutting. The model is simple: design something on your desktop, and upload the file to the website. Software there will examine the file and make sure it's produceable, then guide you through choices on how you want it made. If it's a 2-D image, it can be laser-cut in a range of materials, from plastics of various sorts to woods and even thin aluminum. If it's a 3-D image, it can be 3-D printed or CNC cut in an even wider range of materials. You can design and make something as small as a ring and as large as a table, and if you've made a mistake in your file (which I invariably do), either the software or a human will help you fix it.

As with a photo service, you can also choose to share the files publicly and let others order copies for themselves. You can even create a simple "storefront" in which you get a cut of the revenues anytime someone makes something that you uploaded.

Ponoko doesn't own most of its production machinery. Instead, it's just a software layer between consumers and fabrication shops with spare capacity. The Ponoko website does the tricky work of coaching potentially inexperienced Makers into creating the design files and uploading them in a form that machines can understand. It recommends materials, calculates pricing, and handles the transaction. Then it sends the files to fabrication houses that don't have to deal directly with consumers.

Shapeways does the same for 3-D printing, with a dazzling range of materials that extend from the usual plastics and resins to titanium, glass, and even stainless steel. Costs are calculated based on the materials chosen and volume needed. Something the size of a toy soldier might cost fifteen dollars in plastic, while bigger metal items could run fifty dollars or more. Objects can be printed in monochrome or full color.

Similar services exist for electronics (printed circuit boards), fabrics, and even ceramics. Meanwhile, the grandfather of them all is the Lego company, whose Lego Digital Designer CAD program for kids lets them do exactly the same thing with Lego bricks, creating a design onscreen, then uploading it to the service to be turned into a custom kit that is shipped back to them, looking just like an official Lego kit. Then, if others buy it, the designer will get a cut of the revenues.

What all these services offer, from the machine shops of MFG .com, the low-cost factories of Alibaba, or the one-off digital fabrication of Ponoko and Shapeways, is the ability to make things from your desktop without your having any tools of your own or stepping into a factory. In a sense, global manufacturing has become scale-agnostic. Once factories only worked for the biggest companies with the biggest orders. Now many of them will work at any volume. Smaller batches mean higher prices, of course, but if you're just making a few of something the cost difference may matter little compared to the ability to do it all. The world's supply chains have finally become "impedance matched" to the individual. Anyone can now make anything.

Soon this smart fabrication software will be built right into the CAD programs themselves, such as Autodesk's 123D. Just as you can choose "Print" from your word processor menu, you will soon be able to pick "Make" from your CAD program's menus. What's more, you will be able to choose whether to make "locally," on your own desktop fabricator if you have one (a 3-D printer, CNC machine, or laser cutter), or "globally" in the cloud using one of these services. The software will help you choose whether to use 2-D or 3-D methods, and which materials to pick based on their properties and cost. The final barrier against entry to mass fabrication will have fallen. We will all just be a menu click away from getting factories to work for us. What do you want to make today?

Chapter 13

DIY Biology

**The ultimate Maker dream is programmable matter.
Nature already works that way.**

Laser cutters, 3-D printers, and CNC mills are cool, but like all desktop manufacturing machines they are limited in what they can make, both in materials and complexity. You're not going to make your lunch with one, or even your next pair of shoes. For that you'll need a full-on Universal Fabricator. Just like the Star Trek Replicator, it's a machine that can make almost anything on command. Too bad it's still fictional.

The idea has fired the imagination of science fiction writers for decades. In his novel, *The Diamond Age*, Neal Stephenson imagines an entire society transformed by "matter compilers" that can make whatever you need, rendering scarcity obsolete.

In the beginning was an empty chamber, a diamond hemisphere, glowing with dim red light. In the center of the floor slab, one could see a naked cross-section of an eight-centimeter Feed, a central vacuum pipe surrounded by a collection of smaller lines, each a bundle of microscopic conveyor belts carrying nano-mechanical building blocks—individual atoms, or scores of them linked together in handy modules.

The matter compiler was a machine that sat at the terminus of a Feed and, following a program, plucked molecules from the conveyors one at a time and assembled them into more complicated structures.[52]

That's sci-fi, but something similar is not impossible. MIT professor Neil Gershenfeld thinks it's just twenty or thirty years away.

How will we get there? The path, Gershenfeld argues, will not be simply making 3-D printers and other CNC machines faster and more precise. The problem with those approaches, he says, is that they just "smoosh stuff around." They may squirt it or cut it or heat it, but they're just moving material or changing its state (hardening it). The material itself has no intelligence or sense of what it's supposed to be. Your fabrication machine has to do all the work; the material isn't "helping."

Contrast that with simple Lego blocks. When a child plays with Lego, the blocks correct the child's mistakes—they fit together only if they're lined up right. The larger Duplo blocks guide the child to the correct orientation with beveled edges that exert a force to rotate the parts in the right direction to fit when they're pushed together. The blocks themselves provide a coordinate system—the Lego grid. And when you're done with the blocks, you don't throw them away. You disassemble them and use them to build something else, making them the ultimate recyclable material.

Programmable matter

In a sense, even Lego blocks are "intelligent matter." They carry with them their own rules of assembly and have preassigned functions, such as hinges and wheels.

Sounds crazy? It's not—it's already all around you. That's the way nature works. Crystals, after all, are made of atoms self-assembling into incredibly complex structures, from snowflakes to diamonds. Your own body is made up of proteins assembled under the instruction of your DNA/RNA from amino acids, which themselves are made up of self-assembled atoms. Biology is the original factory.

"Intelligent materials" describe some of the basic building blocks of life. Gershenfeld's favorite example among those are the ribosomes found in your cells. A ribosome is a protein that makes proteins—a biological machine that makes other biological machines. But as Gershenfeld sees it, it's a model of an advanced fabricator.

In your cells, genes coded in DNA are translated into RNA, a sort of mirror image. The ribosome is the "organelle" that reads the RNA and follows that code to assemble amino acids to make up specialized proteins. Once made, those proteins automatically fold into complex shapes, driven by no more than the electrical charges and the attractive and repulsive forces that come from their atomic bonds. Those shapes, self-assembled in the billions, make up the structural elements of your body, from cell walls to bones.

This is a case of a *one-dimensional code* (DNA, four chemical "letters" in different combinations, strung together in a long monodimensional chain) creating a *three-dimensional object* (proteins). Because the materials DNA is working with—first RNA, then ribosomes, then proteins—are not just smooshed together but instead have their own chemical and structure rules and logic, a little information can create incredible complexity. Ribosomes are, Gershenfeld says, "programmable matter." In this case, our DNA programs them. But the same principle could apply to anything.

In Gershenfeld's MIT lab, students have taken some baby steps toward that, with tiny electronic components that can be plugged together and automatically make the right connections. But researchers elsewhere have taken the concept even further. The most promising programmable matter is DNA itself.

The new field of "structural DNA" uses the material not as a genetic code, but as the building material itself, with no biological function. Some sixty labs around the world today are now working on this, and researchers can synthesize strands of DNA that will form squares, triangles, and other polygons.[53] Some of the structures are made by "tiling" many two-dimensional DNA shapes into one sheet. Others program the DNA to fold into three-dimensional shapes, a process called "DNA origami."

Three-dimensional DNA structures can be programmed to assemble into "scaffolding," making structures like a box. Other sequences can be programmed to respond to a chemical stimulus to open, making a door. The idea is that a drug could be placed in a structural DNA box with the door shut and transported by the body to a place where the drug was needed. Then a chemical trigger could be sent to open the door, and the drug would flow out, precisely where intended.

We're a long way from such programmable nanomachines creating large-scale objects of any material. For one thing, DNA is not very rigid, so researchers have experimented with bonding other materials, such as gold nanoparticles, to DNA to strengthen it. Even then, they haven't made anything big enough to be seen without a microscope. Other researchers have experimented with doing similar things with special polymers and other chemical compounds, which have advantages in stiffness but are harder than DNA to program.

It's all been mostly proof-of-concept research so far. But the fact that it works at all suggests that programmable matter at a macro scale is not impossible—maybe even, as Gershenfeld predicts, just a generation away. And it's already got its own Maker Movement.

Making with DNA

Just before midnight on an April Friday in 1983, Kary Mullis, a chemist with a streak of surf bum, was driving along California's Pacific Coast Highway 128 between Cloverdale and Booneville when he had an idea that would eventually win him the Nobel Prize. At that time, one of the biggest problems in genetics was that there was never enough DNA to study, and what DNA could be found was often contaminated.

As he drove, Mullis was noodling over various ways to analyze mutations in DNA when he realized that he had stumbled on a way to reproduce any DNA region with the use of a special bacterial enzyme called DNA polymerase and a process of applying cycles of heat. Others had thought of using the polymerase for copying DNA sections of interest, but Mullis realized that heat cycles would lead to a chain reaction by which each cycle doubled the number of copies, quickly reaching the millions.

In combination with a version of that enzyme derived from bacteria called extremophiles that live in hot springs and are heat-resistant, this led to the automated process of copying DNA that created the modern genetics research industry. Called polymerase chain reaction (PCR), it won Mullis the 1993 Nobel Prize in Chemistry.

Today, PCR machines, also known as thermal cyclers, are a staple in any genetics lab. Once costing nearly $100,000 each, they can now be found commercially for as little as $5,000. PCR is one of the miracles of the genetics revolution and a cornerstone of the New Biology.

But even $5,000 is too expensive. What if you want to use them powered by batteries in Africa? How about with kids in the classroom? What if you wanted to experiment with the machines themselves, not just what went into them?

Josh Perfetto, a young Californian researcher, wanted all of that. And he wanted it to be open and useable by all. So, because he could, he created OpenPCR, an open-hardware thermocycler. It's a lunchbox-sized plywood container with a small LCD screen on top. Inside is an Arduino processor board, a power supply, a container for the DNA and enzymatic bath, and some heater coils. It costs $599, or about a tenth of what a commercial thermocycler does. And it's open source, so you can modify it anyway you want.

Perfetto is part of the "DIYbio" community, which is a small off-shoot of the Maker Movement. "Biohackers" are launching shared science workspaces, like the makerspaces of the hardware world, including Biocurious in Silicon Valley and Genspace in New York City. So far they're still not much more advanced than a typical college biology lab, but they're reaching new audiences with attention-getting (and educational) projects such as DNA fingerprinting sushi sold in local restaurants to see if it's really what it's advertised to be.

So far the DIYbio Movement has been less about doing new science than about democratizing the tools of science. Too much lab equipment is expensive, proprietary, hard to use, and locked down. So, like Perfetto, the biohackers are open sourcing the tools, one by one.

For instance, a lab centrifuge, which spins test tubes to separate heavier materials from lighter ones in solution, can cost thousands of dollars. But it's really just an electric motor with a speed controller and spindle to hold the test tubes. Enter "DremelFuge," a free design for a 3-D printable head that can be mounted on a Dremel rotary tool that can be found in any hardware store—total cost less than $100. Designed by Cathal Garvey, a biologist in Cork, Ireland, it can spin tubes at 33,000 rpm, which can produce up to 51,000 Gs (professional centrifuges typically max out around 24,000 Gs).

DIYbio teams have also created projects from an open-source magnetic stirrer (a way to stir fluids without any seals that might leak) to an open algal bioreactor, which is a way to cultivate algae to create biofuels, animal feeds, or to absorb pollution.

Right now the biohackers are mostly just reinventing the wheel, creating DIY versions of equipment and techniques already found in standard professional and academic labs. They're making lab equipment cheaper, more accessible, and modifiable, but what they're producing with those tools is pretty much standard lab biology. Elsewhere, however, the DIY credo is taking a different twist. Underground synthetic chemists are creating variations of illegal drugs that have the same effect but are chemically different enough to be legal. In a game of cat-and-mouse with the regulators, they have learned to invent new compounds in their DIY labs faster than the regulators can identify and ban them. Synthetic powders that have similar properties to the THC in marijuana are sold legally in head shops in the United States, despite evidence that they can cause more harm than real marijuana.

And that's just chemistry. What happens when the tools get powerful enough to extend to biology and genetics? Today we can amplify and identify DNA at the kitchen table. Tomorrow we will be able to sequence it, too. But after that comes synthesizing it, modifying it, and the rest of genetic engineering. The day when only a small number of professional labs can do this, checking and screening every request that comes in, will soon end. At that point, people will start hacking life. We've been doing that for thousands of years with cross-breeding and agricultural genetics, but that was always within the bounds of nature. But in the lab, there are fewer such bounds. And the DIYbio Movement aims to create countless new labs. Why should trained scientists have all the fun?

Epilogue: The New Shape of the Industrial World

The West can rise again.

What kind of economic future does the rise of the Maker Movement predict?

Is it one where Western countries like the United States regain their lost manufacturing might, but rather than with a few big industrial giants, they spawn thousands of smaller firms picking off niche markets?

Remember that line from Cory Doctorow's book:

> The days of companies with names like "General Electric" and "General Mills" and "General Motors" are over. The money on the table is like krill: a billion little entrepreneurial opportunities that can be discovered and exploited by smart, creative people.

Call this the commercial Web model, one defined by low barriers to entry, rapid innovation, and intense entrepreneurship. Is this the future of manufacturing, too?

Or could it become more like the *real* Web, where the majority of content is created by amateurs, without any intention of creating a business or making money at all?

This second option is a future where the Maker Movement is more about self-sufficiency—making stuff for our own use—than it is about building businesses. It is one that hews even closer to the original ideals of the Homebrew Computer Club and *The Whole Earth Catalog*. The idea, then, was not to create big companies, but rather to *free ourselves from big companies*.

Every time I download some design from the Web and print something on my MakerBot without going to a store or otherwise engaging in any commercial transaction at all, I wonder how long it will take before more of the world of atoms becomes free, like most of the world of bits already has. (I wrote a book about this economic model, too, which now hardly needs explaining as we are awash in free digital goods.)[54]

Take, for instance, Open Source Ecology, which is an online community creating a "Global Village Construction Set." These are open-source designs for the fifty machines necessary to "build a small civilization with modern comforts," ranging from a small sawmill to a micro-combine for harvesting.

This hearkens back to the Israeli kibbutz model of self-sufficiency, which was forged in a period of need and philosophical belief in collective action, or to Gandhi's model of village industrial independence in India. Of course we're not all going to grow our own food or easily give up the virtues of a well-stocked shopping mall. But in a future where more things can be fabricated on demand, as opposed to manufactured, shipped, stored, and sold, you can see the opportunity for an industrial economy that is less driven by commercial interests and more by social ones, just as open-source software already is.

Which one of these manufacturing futures is most likely for the West?

My money is on the first model: something closer to today's commercial Web—ever-accelerating entrepreneurship and innovation with ever-dropping barriers to entry. In this future, the pendulum of manufacturing will swing back to the most nimble developed countries, despite their relatively expensive labor. Globalization and communications flattened the world once, drawing manufacturing to low-cost labor in the developing world, a process first observed in the nineteenth century by David Ricardo as the triumph of "comparative advantage."

Now we are flattening it again, but along a different dimension. Thanks to automation, labor costs are a small and shrinking fraction of the cost of making something. For electronics, they can be just a few percent. At that point, other factors, from transportation costs to time, start to matter more.

For example, the 3D Robotics factory in San Diego buys its electronics manufacturing equipment and components for essentially the same price our Chinese competitors do. We pay our workers better, but the gap is shrinking: around fifteen dollars an hour in San Diego ($2,400 a month) compared with around $400 a month at Foxconn, the huge manufacturing company in China that makes the iPhone, iPad, and electronics for many other leading companies. Because of competition, rising skills, and pressure from labor activists, the salaries in Shenzhen have risen 50 percent in the past five years, while manufacturing salaries have remained close to flat in the West.

In our new 3D Robotics factory in Tijuana, twenty minutes away from our factory in San Diego, the salaries are about half the U.S. rate ($1,200 a month), which is just three times the price of China. For one of our products, such as a $200 autopilot board, the difference in labor costs between making it in Mexico and making it in China amounts to less than a dollar, or about one percent of the product's cost (and half a percent of its retail price). Other costs, such as rent and electricity, are even closer to Chinese levels.

In short, for products that can be made robotically, which is more and more of them, the usual global economic calculus of labor arbitrage is becoming less and less important. Even Chinese firms are moving toward more robotic production, not just to insulate firms from rising salary pressure, but also to avoid the labor condition controversies that dogged Foxconn and Apple for the past few years. Not everything can be automated, of course, and there is a still a lot of handwork in your iPad. But industrial robots are getting cheaper and better all the time, while humans are getting more expensive.

So the decision on where to make things has become less about salaries. Yet China still has a sizable advantage in everything from electronics to toys and textiles, as the labels on your clothes and gadgets prove. Why? Peerless supply chains. Although we do our assembly in the United States and Mexico, the components still come from China and we have to wait for them or stockpile more than we need at any one time, costing us money and limiting our flexibility. In Shenzhen, where all these parts are made, you can order what you need from a neighboring supplier and have it delivered in a few hours. We have to order with weeks of advance notice. Likewise, our plastic injection-molding is done in China because U.S. and Mexican companies don't have the volume to compete on price.

At this point, you can start to see the shape of the twenty-first-century manufacturing economy.

On the product-development side, the Maker Movement tilts the balance toward the cultures with the best innovation model, not the cheapest labor. Societies that have embraced "co-creation," or community-based development, win. They are unbeatable for finding and harnessing the best talent and more motivated people in any domain. Look for those countries where the most vibrant Web communities flourish and the most innovative Web companies grow. Those are the values that predict success in any twenty-first-century market.

On the manufacturing side, the spread and sophistication of automation will increasingly level the playing field between East and West, as will the growing direct and indirect costs of long and brittle supply chains. Every time the price of diesel fuel goes up, so does the price of sending a container from China. Volcanoes in Iceland and pirates off the coast of Somalia—these are among the risk factors in a global supply chain, and are arguments for making goods closer to their point of consumption. We live in an increasingly volatile and unpredictable world, and everything from political uncertainty to currency fluctuations can erase the cost advantages of offshoring in a flash.

But don't think that this means a return to the simple glory days of Detroit or a day when factory jobs were a safe path to the middle class. Instead, it predicts that the Web model really will hold sway: a fully distributed digital marketplace where good ideas can come from anywhere and take the world by storm. Think of the rise of Angry Birds (created in Finland) and Pinterest (founded in Iowa) rather than domination by the traditional manufacturing centers and companies of the twentieth century.

General Motors and General Electric aren't disappearing, but then again, neither did AT&T and BT as the Web arose. As with the Long Tail, the new era will not mark the end of the blockbuster, but the end of the *monopoly of the blockbuster*. So, too, for manufacturing. What we will see is simply *more*. More innovation, in more places, from more people, focused on more narrow niches. Collectively, all these new producers will reinvent the industrial economy, often with just a few thousand units at a time—but exactly the right products for an increasingly discriminating consumer. For every Foxconn with a half-million employees making mass-market goods, there will be thousands of new companies with just a few targeted niches. Together they will reshape the world of making.

Welcome to the Long Tail of things.

The 21st-Century Workshop

How to become a Digital Maker.

I hope at this point in the book you've been inspired enough to want to try it yourself. How to get started? The answer, of course, depends on what you want to do, and there are as many answers to that as there are people asking the question. Making can be as simple as kitchen-table crafting and as complicated as a machine shop. There are many terrific Maker resources out there to guide you, including the wonderful *Make* magazine, websites such as Instructables, and countless crafting magazines and websites.

But the theme of this book is the power of digital tools, the desktop fabrications revolution. So in this appendix, I'll give a guide to starting with that, using the best recommended tools as of this writing.

Most of this is based on personal experience. I've got a little workshop in our basement, and it's outfitted with the sorts of tools I need for projects with the kids, some robotics and electronics and generally experimenting with digital fabrication.[55] It includes elements of those listed here, but everything in the list is something I've got some personal experience with, and can recommend.

Getting started with CAD

Why? All digital design revolves around software. Whether you're downloading designs or creating them from scratch, you'll typically need to use some sort of desktop authoring program to work with the design onscreen.

Think of CAD as the word processor of fabrication. It's just a way to get your ideas onscreen and edit them. CAD programs range from the free and relatively easy Google SketchUp to complex multithousand-dollar packages such as Solidworks and AutoCAD used by engineers and architects.

There are also all sorts of specialty CAD programs, such as those that allow you to design printed circuit boards for electronics (such as the Cadsoft Eagle program) or even those that let you design biological molecules. But in this appendix I'll just focus on those designed to create the sort of objects that you can fabricate on a 3-D printer, CNC machine, or laser cutter.

The first distinction to make is between 2-D design and 3-D design. Some desktop fabrication machines, such as simple laser cutters, just cut flat materials like a pair of scissors. That makes them 2-D machines, and as a result all you need to control them is a 2-D outline image. That's easy to create in any "vector" drawing program, such as Adobe Illustrator or CorelDRAW.

Such drawing programs are similar to the simple "paint" programs that come free with Windows and the Mac, but with the difference that each line and shape is an "object" that can be independently edited at any time, moving them, stretching them, or deleting them. When you're done, those lines will be interpreted as "toolpaths" for the laser head in the laser cutter or simple CNC router: they tell the head where to go and cut. They're easy to use and tend to involve little more than selecting standard shapes such as circles and rectangles and simply stretching and combining them to get the shape you want to cut out of sheets of plywood, plastic, or thin metal.

Recommended 2-D drawing programs
- **Free option:** *Inkscape (Windows and Mac)*
- **Paid option:** *Adobe Illustrator (Windows and Mac)*

For more complex objects that will be printed on a 3-D printer or milled on a 3-axis CNC machine, you'll need a 3-D drawing program. Because you're essentially sculpting a 3-D object on a 2-D computer screen, these require a little more mental and visual gymnastics. These are the same sorts of tools used by Hollywood and the video-game industry to design computer graphic animations, but in this case you'll be making objects that can be fabricated in real materials. It's basically the same process, but you'll need to be more careful to ensure that parts that are supposed to actually touch each other do so and that there are no gaps that will confuse your 3-D printer or CNC machine (the dreaded "leaky mesh" problem).

Typically, in 3-D CAD programs you start with simply placing "geometric primitives" such as rectangles and circles on the screen and then "extruding" them, pulling them out into 3-D objects that you can then manipulate further. Combine enough such elements and you can design anything, from the most complex machinery to the human form.

Recommended 3-D drawing programs
- **Free options:** *Google SketchUp (Windows and Mac), Autodesk 123D (Windows), TinkerCAD (Web)*
- **Paid option:** *Solidworks (Windows and Mac)*

Getting started with 3-D printing

Why? If you can imagine it, you can make it. A 3-D printer is the ultimate prototyping tool, the fastest way to turn something from bits on the screen to atoms in your hand. But remember that the ones designed for home use are still pretty crude. The things you make may work, but they won't be pretty.

Just a few years ago, 3-D printers cost tens of thousands of dollars and were used only by professionals. Now, thanks to a wave of open-source projects, starting with the RepRap printer and then the popular MakerBot, 3-D printing has fallen below $1,000 and printers are found in schools, homes, and countless makerspaces.

All of the 3-D printers around $1,000 create objects with layers of melted ABS plastic, which is fed in spools of filament in varying colors. This can create a tough, flexible material, but it's limited to about half a millimeter in resolution. That can create good-looking prints, but nobody will confuse them with the smooth and seamless quality of professional 3-D printers that use lasers instead.

I've recommended the MakerBot Replicator below, which is what I currently use, but this is a fast-moving field and there will no doubt be cheaper and better options by the time you read this (some of them no doubt brought to you by MakerBot Industries).

Think of these early consumer-grade 3-D printers as the dot-matrix printers of their day: great for drafts and prototypes, but you'll still probably want to use a professional printing service such as Shapeways or Ponoko for the final version.

Recommended 3-D printing solutions
- Printers: *MakerBot Replicator (best community), Ultimaker (bigger, faster, more expensive)*
- Services: *Shapeways, Ponoko*

Getting started with 3-D scanning

Why? Properly set up, 3-D scanners can digitize the world faster than CAD software.

One of the hardest parts of working with 3-D objects is creating them in the first place. You can jump-start that process by scanning an existing object and then modifying it in a CAD program. Such 3-D scanning is called "reality capture," and is typically done with a special scanner or just many shots from a regular camera, all stitched together with smart software.

Professional-grade 3-D scanners cost thousands of dollars, but you can get surprisingly good results with cheap or even free products that use a digital camera if you're careful with your lighting.

The easiest option is to use a good digital camera to take lots of pictures from all angles of your object and then use the free Autodesk 123D Catch software to upload it to the cloud to be stitched together and sent back as a "point cloud," which can be rotated and manipulated. This works best for objects you can photograph from all sides in natural light against relatively varied backgrounds, such as a chair or even a room.

For smaller objects, you'll do better with a stand-alone 3-D scanner that combines a camera with a "structured light" projector, which shines a known pattern on the object to reveal all of its ins and outs. If you use an inexpensive webcam-based scanner such as the MakerBot one listed below, it will need a good bit of software cleanup afterward. To avoid that, you'll need a professional scanner, which will cost thousands of dollars. A better solution if you want to scan relatively small objects and not often is to use a scanning service to which you can send the object.

Someday 3-D scanners will be as ubiquitous as the 2-D flatbed scanner that's probably built into your current desktop all-in-one paper printer today. But for now they're still a somewhat fiddly technology. Capturing the image is easy enough, but using the software to clean it up so you can work with it onscreen is still something of an acquired skill.

Recommended 3-D scanning solutions
- Software: *Free Autodesk 123D Catch (iPad; Windows)*
- Hardware: *MakerBot 3-D scanner (requires a webcam and pico projector). Use the free Meshlab software to clean up the image.*

Getting started with laser cutting

Why? Anybody can make something cool with a laser cutter, from jewelry to a bird feeder or even furniture. If you can draw it on paper, you can make it.

The easiest digital fabrication you can do is to use a laser cutter. All you need is a 2-D drawing (see page 233) or a 3-D drawing that is automatically "sliced" into 2-D layers with software such as the free Autodesk 123D Make app. The machine does all the rest of the work, tracing along that line with a high-powered laser that can cut through wood, plastic, and even thin metal.

Although a laser cutter is simple to use, it's probably the least necessary tool in a home workshop. That's because it's so easy to upload files to a service bureau and get them made for you cheaply in a few days. Unlike more complex 3-D fabrication, with laser cutting it's pretty easy to predict what you're going to get, sight unseen, and the service bureau websites will help you choose the right material to use. Laser cutters also tend to be pretty expensive for a home workshop, with the cheapest ones that cut any decent thickness of material costing around $2,000. And they can spew out some unpleasant fumes when they're cutting plastic, so you'll need a good ventilation system.

All in all, I recommend that you either do your laser cutting at a local makerspace such as TechShop, or send it to a service bureau that can also source the raw material for you cheaply.

Recommended laser-cutting solutions
- Service bureau: *Ponoko.com*
- Software: *Autodesk 123D Make (Mac only at the time of this writing, but Windows in the works)*

Getting started with CNC machines

Why? They're relatively easy to use, can fabricate in almost any material, and come in desktop versions cheaper and smaller than a laser cutter.

All 3-D printers are an "additive" technology, which means that they lay down layers of some material to make an object. That means you're typically limited to the kinds of materials that you can melt, and for low-cost printers that means plastics. If you want to make things inexpensively out of other materials, such as wood or metal, you'll be better-off with a "subtractive" technology, which means rotating, grinding, or cutting heads that can remove material. So, rather than lay down material where the thing "is," you remove material where it "isn't."

The simplest CNC machines are just a holder for a rotary power tool like a Dremel that can move in three directions (x, y, and z) on rods in response to computer control. Software on a desktop computer determines the tool paths for a 3-D object and moves the spinning power tool head accordingly. A milling bit grinds away material until just the desired object is left. More expensive ones use specialized power tool heads and bits that can grind, cut, or even polish.

Unlike a laser cutter, CNC routers and mills can cut precisely in three dimensions, so they can make complex shapes with layers. The more expensive ones have four or even five axes of movement, so the head can twist around to get into nooks and crannies in the object.

Beginners can use a CNC machine as they might use a manual jigsaw, precisely cutting out patterns in flat material such as plywood. More advanced users can extend that to 3-D, milling more complex objects ranging from aluminum molds for plastic injection-molding to metal robot parts.

Recommended CNC Solutions
- Hobby-sized (Dremel tool): *MyDIYCNC*
- Semi-pro: *ShopBot Desktop*

Getting started with electronics

Why? A big part of the Maker Movement is making physical objects smarter—giving them sensors, making them programmable, and connecting them to the Web. This is the emerging "Internet of Things," and it starts with simple electronics such as the Arduino physical computing board.

All you really need to get started with digital electronics is an Arduino starter kit, a multimeter, and a decent soldering iron. Depending on what you want to do, you may want to try various sensors or actuators such as servos or motors. There has never been a better time to find what you need, and companies such as Sparkfun and Adafruit offer not only all the parts you'll need, but also tutorials, great product instructions, and large communities to help. It really is the second golden age of electronics tinkering! (The first was the post–World War II ham radio days which culminated in the Heathkit era in the 1970s. Then inscrutable microchips ruined tinkering for a generation or so until open-source hardware brought it back in the last few years.)

If you want to take it further, you can get a digital logic analyzer, a USB oscilloscope, and a fancy solder rework station. But for starting, the items listed below will take you further than you may have thought possible.

Recommended electronics gear
- Starter kit: *Adafruit budget Arduino kit*
- Soldering iron: *Weller WES51 soldering station*
- Multimeter: *Sparkfun digital multimeter*

Acknowledgments

This book and my introduction to the world of Makers began with a 2007 weekend effort to get the kids interested in programming and robotics with Lego Mindstorms. Although that didn't work as well as I had intended for them, it did wonders for me. It ultimately led to the "Lego UAV" project and my descent into drone madness, so my first thanks go to Lego for its brilliant robotics construction set—fun for all ages!

Thanks also to my wife and five kids for indulging these obsessions for so long, and letting me add a whole new floor to our house for my workshop (the kids think that part of it is their TV room, but we'll see how long that lasts). And also for playing along with my obsessions in good humor, with each child finding some project that allowed us to explore the Maker world together.

My inspirations and guides in this world are many. Dale Dougherty of O'Reilly, who started *Make* magazine and the Maker Faire, was among the first to capture the growing movement, and Mark Frauenfelder's enthusiasm drove it forward. Cory Doctorow inspired us all with his visions of how far it could go. So did countless blogs and Web communities, from Hackaday and Makezine to Instructables, Kickstarter, Etsy, and Quirky.

In the open-hardware movement, my chief guides have been Massimo Banzi of the Arduino project, Nathan Siedle of Sparkfun, Limor Fried and Phillip Torrone of Adafruit, Bre Bettis of MakerBot Industries, and Jay Rogers of Local Motors.

Among the toolmakers, Carl Bass, the CEO of Autodesk, and his team have been a huge inspiration and source of support, as were Jim

Newton and Mark Hatch of TechShop, and Derek Elley and David ten Have of Ponoko.

At *Wired*, where my passions are indulged and even sometimes encouraged, my eternal thanks go to the crack team who find ways to bring these worlds to our pages (print, tablet, and Web), especially Thomas Goetz, who also writes books while running a magazine and having a family—no small task, as I can attest. Thanks also to Shoshona Berger, who has found remarkably creative and energetic ways to turn the Maker Movement into a mainstream media site with Wired Design. Note that this book started as an article in *Wired* and includes some text from that, as well as excerpts from my writing on this subject elsewhere.

Finally, my greatest thanks (and dedication of the book) go to my mother, Carlotta Anderson, for having the wisdom to see that what her father was doing in his workshop was more than just tinkering and might inspire her kid to someday do more than just tinker himself. Although I didn't know it at the time, those summers my parents sent me to learn from my grandfather would end up changing my life. It just took thirty years—and a new era of technology—for the seeds to sprout. Thanks, Mom, for keeping Grandpa's tools, patents, and drawings, so you could someday pass them on to me. They mean more to me now than I could have ever known.

Notes

1. http://www.citibank.com/transactionservices/home/docs/the_new
   _digital_economy.pdf
2. One of them was called REM, and started at the same time as the
   "other" REM. For an amusing story on how the two bands resolved
   who would get the name (short form: a battle of the bands, which we resoundingly lost. We emerged called "Egoslavia"), you can read my post
   here: http://www.longtail.com/the_long_tail/2006/07/my_new_wave
   _hai.html.
3. Cory Doctorow, *Makers* (New York: Tor Books, 2009).
4. http://www.engadget.com/2011/11/12/shanghai-science-and-tech
   nology-commission-proposes-100-innovat/
5. http://www.auctionbytes.com/cab/abn/y12/m02/i07/s02
6. http://online.wsj.com/article/SB1000142405270230422380457644
   042827007476.html
7. http://makerspace.com/2012/01/16/darpa-mentor-award-to-bring
   -making-to-education/#more-43
8. As you'll see throughout this book, I'm not just an observer in this
   movement, I'm also a participant. Along with 3D Robotics, which I
   cofounded, I've worked with Autodesk, Ponoko, and others to help
   steer this evolution, sometimes as a member of the company's advisory board. This book is built on my experience at the front lines, and
   although I've endeavored to use journalistic standards of reporting,
   research, and fact-checking, I don't pretend to be impartial. This is a
   future I believe in, and I'm one of many working to build it.
9. http://www.kickstarter.com/blog/2011-the-stats
10. http://www.businessinsider.com/kickstarter-on-track-to-generate
    -300-million-in-2012-funding-2012-4

11. http://www.crunchbase.com/

12. http://www.wired.com/epicenter/2011/10/jobs/all/1

13. Walter Isaacson, *Steve Jobs* (New York: Simon & Schuster, 2011, Kindle Edition), Kindle locations 1252–1264.

14. William Rosen, *The Most Powerful Idea in the World* (New York: Random House, 2010), 214.

15. http://apps.business.ualberta.ca/rfield/lifeexpectancy.htm

16. François Crouzet, *The Industrial Revolution in National Context: Europe and the USA* (Cambridge, UK: Cambridge University Press, 1996).

17. http://www.ribbonfarm.com/2011/06/08/a-brief-history-of-the-corporation-1600-to-2100/

18. http://www.historywithmrgreen.com/page7/assets/The%20Industrial%20Revolution%20Cottage%20Industry%20and%twentieth%20Factory%20System.pdf

19. http://www.nytimes.com/1992/03/30/business/plug-is-pulled-on-heathkits-ending-a-do-it-yourself-era.html

20. Michael J. Piore and Charles F. Sabel, *The Second Industrial Divide: Possibilities for Prosperity* (New York: Basic Books, 1984).

21. http://www.nytimes.com/2012/02/19/magazine/adam-davidson-craft-business.html

22. http://www.sciencedirect.com/science/article/pii/S1057740811000829

23. http://mitpress.mit.edu/catalog/item/default.asp?ttype=2&tid=12470

24. Neil Gershenfeld, *Fab: The Coming Revolution on Your Desktop* (New York: Basic Books, 2005).

25. http://www.washingtonpost.com/national/on-innovations/the-past-present-and-future-of-3-d-printing/2011/08/21/gIQAg4fJZJ_story.html

26. http://www.forbes.com/sites/richkarlgaard/2011/06/23/3d-printing-will-revive-american-manufacturing/

27. http://investor.cafepress.com/secfiling.cfm?filingID=1193125-12-135260&CIK=1117733

28. http://www.slideshare.net/adafruit/hope2010-4790096

29. http://www.americanheritage.com/events/articles/Web/20070709-windshield-wiper-robert-kearns.shtml

30. http://www.washingtonpost.com/wp-dyn/articles/a54564-2005
feb25.html

31. http://www.dieselpowermag.com/features/1005dp_local_motors
_rally_fighter/

32. http://www.meyersmanx.com/pdf-files/Meyers_Manxter2.pdf

33. http://www.jstor.org/pss/2626876

34. http://www.econlib.org/library/Essays/hykKnw1.html

35. As an aside, the fact that so much of the team is from Tijuana is worth
noting. What I've learned, now that I've seen the city through the
eyes of its young entrepreneurial generation, is that it's much more
a high-tech manufacturing zone (almost all of the flat-screen TVs
we buy are made there) than the drug battleground or tequila drinking strip cemented in many Americans' imagination. The kids growing up in "TJ" over the past few decades are part of the California
extended tech corridor, which stretches from San Francisco to San
Diego. They have access to all the technology that's twenty minutes
across the border in San Diego, but at a fraction of the cost. Think
of it like the Hong Kong/Shenzen nexus: lower-cost labor across the
border, but equal skills.

36. http://www.aspeninstitute.org/sites/default/files/content/docs/pubs/
The_Future_of_Work.pdf

37. http://hbr.org/hbr-main/resources/pdfs/comm/fmglobal/restoring
-american-competitiveness.pdf

38. Kenneth L. Kraemer, Greg Linden, and Jason Dedrick, "Capturing Value in Global Networks: Apple's iPad and iPhone, http://pcic
.merage.uci.edu/papers/2011/Value_iPad_iPhone.pdf

39. http://www.bcg.com/media/PressReleaseDetails.aspx?id=tcm:12
-75973

40. http://money.cnn.com/magazines/fortune/global500/2011/performers/
companies/biggest/

41. http://curiouscapitalist.blogs.time.com/2011/01/11/is-the-iphone
-bad-for-the-american-economy/

42. http://radar.oreilly.com/2011/05/crowdfunding-exemption.html

43. http://www.whitehouse.gov/the-press-office/2012/04/05/president
-obama-sign-jumpstart-our-business-startups-jobs-act

44. http://www.washingtonpost.com/blogs/innovations/post/the

-underground-venture-capital-economy/2010/12/20/gIQAzkRQvJ
_blog.html

45. http://cultureconductor.com/author/sarahdopp/
46. http://www.bbc.com/news/technology-17531736
47. http://www.wired.com/magazine/2011/03/ff_kickstarter/all/1
48. http://www.etsy.com/blog/news/2012/notes-from-chad-funding
    -etsys-future/
49. http://www.slideshare.net/stevekeifer/b2b-emarketplaces-rise-and
    -fall-by-steve-keifer
50. http://ftp.iftf.me/public/IFTF_open_fab_China_conversation.pdf
51. http://www.iftf.org/LightweightInnovation
52. Neal Stephenson, *The Diamond Age: Or, a Young Lady's Illustrated Primer* (New York: Bantam Spectra, 1995).
53. http://mag.uchicago.edu/science-medicine/crystal-method
54. Chris Anderson, *FREE: The Future of a Radical Price* (New York: Hyperion, 2009).
55. Here are the main tools in my workshop:

Hardware:
- First-generation MakerBot Cupcake, upgraded as much as possible
- MakerBot Cyclops 3-D scanner
- MyDIYCNC
- Hitachi desktop bandsaw
- Dremel workstation/drill press
- Weller WES51 soldering station
- Picoscope USB oscilloscope
- Saleae USB logic analyzer
- Velleman Power supply/Multimeter/Soldering station

Software:
- Adobe Illustrator (for laser cutting drawings)
- Autodesk 123D (for 3-D)
- Cadsoft Eagle (for PCB design)
- Arduino, Notepad++ and TortoiseSVN and TortoiseGIT for version control

Index